Crop Diseases and Their Management

Crop Diseases and Their Management

Editor

Andrick Cuevas

scitus
academics

Crop Diseases and Their Management
Edited by **Andrick Cuevas**

ISBN: 978-1-68117-705-2
Library of Congress Control Number: 2016958739

616, Corporate Way, Suite 2, 4766,
Valley Cottage, NY 10989

www.scitusacademics.com

Printed in United States of America on Acid Free Paper

Preface

Diseases cause huge loss of crop plants. In human being we found diseases like Malaria, typhoid, cholera etc. Likewise plants also get affected by number of disease causing reasons. Crop diseases are as old as agriculture itself and there are many examples in history where people have suffered as a consequence of plant diseases. Diseases cause economic losses to crops by attacking the plant at many different places and stages in its growing cycle. Some will cause damage by interfering with water and mineral absorption from the soil, some will affect photosynthesis by killing the leaves of the plant, some will impair translocation of sugars produced in the photosynthesis to the grain and yet others will completely destroy developing grains.

Crop Diseases and Their Management outlines the general principles of plant pathology, the conditions that favor disease development, how to scout for diseases and how to control them. It is designed to provide information to the readers on all important aspects of plant pathology in a single volume. The information on modern areas like disease diagnosis, disease forecasting, biological control, epidemiology and biotechnology in disease resistance and safe use of pesticides have been covered, giving most recent concepts. It is intended of valuable for plant protection experts, students and researchers, vegetable specialists, horticultural officers, and extension workers.

Table of Contents

Chapter 1

Characterization of *Apple mosaic virus* isolates detected in hazelnut in Poland

Mirosława Cies´lin´ska[1] • Natallia Valasevich[2]

[1]Department of Plant Protection Research Institute of Horticulture Skierniewice Poland

[2]Department of Biotechnology Institute for Fruit Growing Samochvalovichi Belarus

ABSTRACT

Apple mosaic virus (ApMV) infects over 65 plants species of *Rosaceae* family including hazelnut. The virus causes chlorotic or yellow patterns, rings and mosaic symptoms on the leaves and losses in yield of nuts. The occurrence and genetic diversity of ApMV were evaluated in 14 plantations of hazelnut in Poland. A total of 13 of 125 hazelnut samples tested positive for ApMV using ELISA and reverse transcriptase-polymerase chain reaction. Several primer pairs were used for RT-PCR to amplify the coat protein (CP) and movement protein (MP) genes as well as complete RNA-3 of the ApMV. Genetic diversity was observed in full-length sequences of RNA-3 of the ApMV isolates from hazelnut. The similarity of nucleotide sequence of the CP and MP genes of the isolates was 90.3–99.9, and 87–99.7 %, respectively. Phylogenetic analysis showed that ApMV isolates from hazelnut grouped separately from the virus strains found in other plant species.

Keywords: ApMV Coat protein Movement protein Phylogenetic analysis

INTRODUCTION

Apple mosaic virus (ApMV) is a member of the subgroup III of the genus *Ilarvirus,Bromoviridae* family. This virus occurs worldwide and infects a number of woody and herbaceous plants in over 65 species in 19 families including apple, hazelnut, strawberry, apricot, black cherry, almond, red currant, raspberry, rose, birch, chestnut, and hop [4, 10, 14,26, 28]. ApMV is transmitted by vegetative propagation of infected plant material and by mechanical inoculation [5]. No natural vector of the virus is known [20]. ApMV causes either chlorotic or yellow patterns, rings and mosaic observed during mild temperatures on the leaves of hazelnut (*Corylus avellana* L.) or infected plants do not show any symptoms. ApMV causes significant yield reductions in hazelnuts (up to 42 %) and infected trees showed a slight growth reduction [1, 3]. The hazel mosaic symptoms were first reported in Bulgaria [7]. The incidence of ApMV on hazelnut plantations has been reported in Turkey [1], Spain [3], USA [21], and Italy [22]. The previous study using ELISA assay showed that ApMV occurred in hazelnut grown in several plantations mostly in south-eastern Poland [13, 19].

Single-stranded RNA of the ApMV genome consists of three segments: RNA-1, RNA-2, and RNA-3 (3.4, 2.9, 2.0 kb size) [23, 24]. RNA-3 has two open reading frames from which the first encodes the putative movement protein (MP, 861 nt) and the other translates coat protein (CP, 672 nt) by producing a subgenomic mRNA [2]. Molecular diversity of ApMV isolates from different geographical regions and plant species has been revealed on the basis of sequence analysis [11, 14, 15, 17, 18, 27].

The aim of the study was to characterize the molecular properties of ApMV isolates from hazelnut and to evaluate the presence of possible genetic diversity of the virus strains found in Poland based on the MP and CP gene sequences and their relationships with previously studied strains.

MATERIALS AND METHODS

Hazelnut plants growing in 14 commercial plantations as well as in home gardens and natural environment located in eight regions of Poland were surveyed for symptoms of apple mosaic disease in 2010 and 2012. A total of 125 leaf samples (56 in 2010 and 69 in 2012) were collected from symptomatic and asymptomatic hazelnut plants of different cultivars (Table 1). The samples were tested by DAS-ELISA for *Apple mosaic virus* using a commercially available polyclonal antibody kit (Loewe Biochemica, Germany).

The silica capture (SC) method [8] was used for extraction of total nucleic acids from leaf samples in which the virus had been detected by ELISA. One-tube reverse transcription-polymerase chain reaction (RT-PCR) was conducted using SuperScriptIII™ One-Step RT-PCR with Platinum® Taq (Invitrogen). Primer sets ApMVup1061/ApMVd1845 [27] and ApMVup/ApMVd [14] were used to amplify the CP and MP genes of ApMV, respectively. Amplification products were separated by electrophoresis in 1 % agarose gel and, after staining with ethidium bromide, were visualized under UV light.

The complete RNA-3 of ApMV which covered both MP and CP genes was synthesized using two pairs of primers: 3ApMV20F/3ApMV1031R and 3ApMV888F/3ApMV2035R [27]. The resulting amplicons were purified, sequenced and analyzed using Lasergene 7.1 software (DNASTAR). The phylogenetic tree was calculated by the neighbor joining method with the maximum composite likelihood model and 1000 bootstrap replications in MEGA 4.0 program [25].

Table 1. Result of testing of the hazelnut samples for *Apple mosaic virus*

Hazelnut cultivar	Number of surveyed plantations	Number of tested plants	Number of positively tested plants
Catalan	7	23	0
Webb's Prize Cob	5	12	0
Hall's Giant	5	9	0
Merville de Bollwiller	3	5	0
Syrena	3	6	0
Lambert's Red	3	8	0
Trapezunter	2	6	**6**
Barcelona	2	5	0
Cosford	2	3	0
Truchsess Zellernuss	1	2	0
Negret	1	7	**7**
Tonda di Giffoni	1	2	0
Waterloo	1	2	0
Lambert's White	1	3	0
Unknown cultivars	6	33	0

RESULTS

In 2010, yellow patterns and mosaic symptoms were observed on the leaves of hazelnut plants 'Negret' (Fig. 1) growing in one of the surveyed plantations located in south-west Poland. *Apple mosaic virus* was detected by DAS-ELISA only in seven symptomatic plants of 'Negret'. The virus was not detected in any of the other cultivars. During the next inspection conducted 2 years later on the same plantations, similar symptoms were noticed also on 'Trapezunter' hazelnut growing in the same row with 'Negret' symptomatic plants. DAS-ELISA resulted in detection of ApMV in leaf samples not only from previously infected 'Negret' hazelnut but also from five 'Trapezunter' plants in which the virus had not been detected in 2010. A single positive sample of 'Trapezunter' was obtained from another plantation.

Figure 1. Yellow line patterns and rings on the leaf of 'Negret' hazelnut infected with *Apple mosaic virus*

RT-PCR with ApMVup/ApMVd and ApMVup1061/ApMVd1845 pairs of primers specific to MP and CP genes of the virus, respectively, resulted in amplification of the products of expected size (860 and 784 bp, respectively).

Five isolates of the virus from Trapezunter cv. (Trapez A, Trapez K, Trapez K5, Trapez 4, Trapez 6) and three from Negret cv. (Negret, Negret 2, Negret 7) hazelnuts were selected for sequence and phylogenetic analyses. The CP of the hazelnut isolates was 663–672 nucleotides (221–224 amino acids) in length. The ranges of pairwise nucleotide sequence similarity in the CP gene of ApMV isolates from hazelnut were 90.3–99.9 and 85.2–90.8 % when compared to corresponding sequences of the ApMV strains available in EMBL/GenBank database. The sequence of this genome fragment of the hazelnut isolates showed maximum identity with almond isolate from Italy (acc. no. AY054388). Phylogenetic analysis of CP gene showed that hazelnut isolates from Poland grouped separately from the virus strains from the other plant species including apple, pear, *Prunus* sp., rose, hop, *Trebouxia* sp. (lichen symbiotic algae), and elderberry (Fig. 2). They were also genetically different from the hazelnut isolate of ApMV reported from the Czech Republic [11]. The 9-nt-deletion in position 155–163 (after 48 aa position) was found in the CP genes of Negret 2, Negret 7, Trapez 4 and Trapez 6 hazelnut isolates, while 6-nt-deletion in position 155–160 was present the sequences of Trapez A and Trapez K isolates. These deletions resulted in missing of 3 aa and 2 aa, respectively, in the putative translations of this fragment of RNA-3. The other specific fragment of the sequence in the CP region of the all eight analyzed hazelnut isolates was AAT → GGT substitution resulted in Asn → Gly change inside MVCKYCGHT (the zinc-finger) motif.

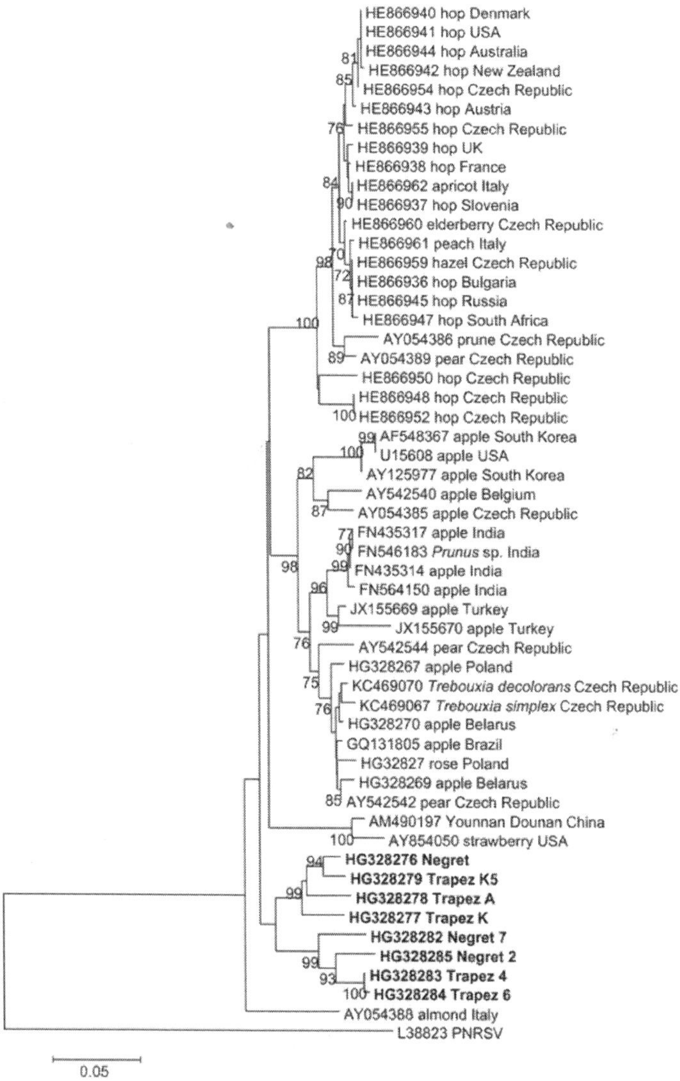

Figure 2. Neighbor-joining analysis showing predicted relationships between *Apple mosaic virus* isolates based on the nucleotide sequences of coat protein (CP) gene fragment. Sequence of *Prunus necrotic ringspot virus*, PNRSV (acc. no. L38823) was used as an outgroup. *Numbers* represent bootstrap values out of 1000 replicates. Bootstrap values below 70 % were collapsed

Sequence of CP gene of isolates Negret, Trapez K, Trapez A, Trapez K5 were deposited in GenBank under accession numbers HG328276–328279, respectively.

The MP gene of the hazelnut isolates was 858 nucleotides (289 amino acids) in length. Multiple alignment of the MP gene sequences showed the heterogeneity of the studied ApMV isolates. The eight hazelnut isolates shared 87.0–99.7 % similarity, while the level of identity of these isolates with apple strains from India, USA, China, Poland and Belarus deposited in GenBank was 86.0–89.4 %. The sequences of MP gene of the hazelnut isolates showed maximum identity with an apple isolate from China (acc. no. AM490197). Phylogenetic analysis of the MP gene showed that Polish hazelnut isolates formed a distinct cluster separated from apple strains from India, Poland, Belarus, Slovakia and USA (Fig. 3). Comparison of the putative amino acid sequences of the MP showed five substitutions which were unique for the Polish hazelnut isolates and several substitutions characteristic for hazelnut isolates and apple strain from China (acc. no. AM490197) only.

The sequences of the MP genome fragments (735 bp) of the isolates from Trapez K5, Trapez A, Trapez K, and Negret have been deposited in GenBank under accession numbers HG328272–328275, respectively. The nearly complete nucleotide sequences of RNA 3 (1934–1939 nt) of four ApMV isolates: Negret 7, Trapez 4, Trapez 6, Negret 2 were deposited in GenBank under accession numbers HG328282–328285, respectively.

DISCUSSION

ApMV was detected in all 160 hazelnut cv. Negret orchards assayed in Catalonia, north-east Spain, in the last decade of twentieth century [5] which indicated that cv. Negret contributed to virus spread. Our study identified the virus also in the Trapezunter cultivar. The presence of ApMV in hazelnut germplasm in Poland from plant material of foreign origin was previously verified by ELISA method, but the virus had not

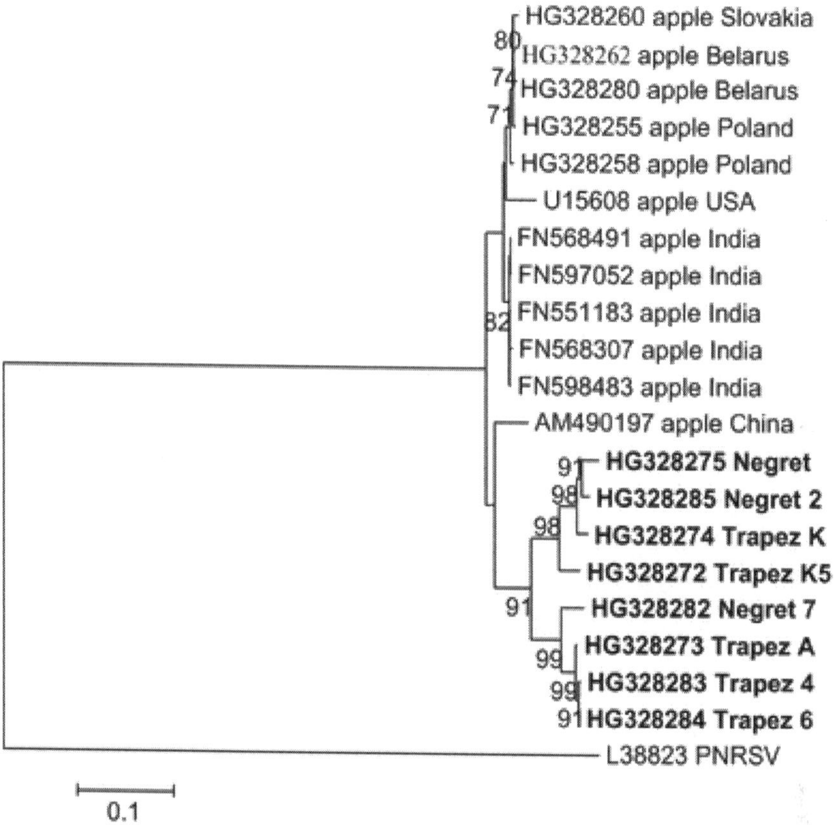

Figure 3. Neighbor-joining analysis showing predicted relationships between *Apple mosaic virus* isolates based on the nucleotide sequences of movement protein (MP) gene fragment. Sequence of *Prunus necrotic ringspot virus*, PNRSV (acc. no. L38823) was used as an outgroup. *Numbers* represent bootstrap values out of 1000 replicates. Bootstrap values below 70 % were collapsed

been detected in Trapezunter cv., while Negret cv. plants were positively tested for ApMV [13]. During our study in 2010, the virus was not found in the samples collected from 'Trapezunter' growing in the same row with 'Negret' trees, while 2 years later its presence was confirmed in plants of the both cultivars. ApMV has been recently detected in naturally infected weeds in hazelnut orchards, and it was reported that the virus could spread via root grafting, weeds or an unidentified arthropod vector [6]. Furthermore, Akbaş and Değirmenci [1] suggested that transmission by foliar or root contact among trees in hazelnut shrubs may be a pathway of ApMV spread in commercial orchards. Foliar contact and root grafting also played role in spreading of this virus in hop plants and apple trees [9, 12, 16].

The sequence analyses of MP and CP genes showed the genetic diversity of the ApMV isolates from hazelnut trees in Poland and allowed to establish their phylogenetic position separately from the other strains of the virus. Variability of the CP gene sequence of the ApMV isolates from different plant species and geographical origin resulted in their classification into separate clusters. There are several CP gene sequence-based ApMV classification systems clustering the virus strains into three [15] or five main subgroups [14]. Recently, Grimová et al. [11] proposed classification of ApMV isolates into two main clusters I and II divided into subclusters. This scheme seems to be not so clear since some virus strains like Italian PAM-ITA from almond (acc. no. AY054388), American NCGR 9026 from strawberry (acc. no. AY854050), and XX1-CHN found in unknown Chinese host (acc. no. AM490197) were not classified to any of these groups. The Polish isolates from hazelnut were also grouped separately from these both clusters on the phylogram. It is in the agreement of the previous studies which showed that there was no correlation between the CP gene sequence variability of the ApMV strains and their geographical origin.

Lee et al. [15] demonstrated that ApMV sequences of different isolates showed maximum variability in the first 300 nucleotides (100 amino acids) of the CP region. The gene sequences of isolates from different plant hosts including hop, peach, elderberry, apricot and hazelnut with the 15-nt-deletion after position 141 resulting in missing of 5 aa [17] were classified by Grimová et al. [11] in Ia subcluster. This deletion does not occur in the CP gene of Polish hazelnut isolates and in the sequence of the strains classified by Grimová et al. [11] into subclusters Ib, IIa and IIb. However, partial deletions of 9 nt and 6 nt were present after position 154 in the sequences of this region of hazelnut isolates from Poland. The 9-nt-deletion was also identified after position 150 in the sequence of PAM-ITA isolate from almond (acc. no. AY54388).

Moreover, in the sequence of hazelnut isolates, Asn → Gly change inside the zinc-finger motif was noticed as a consequence of AAT → GGT substitution. Grimová et al. [11] reported that the position Gly7 in the context MVCKYCGHT is characteristic for all the sequences from cluster I as well as for PAM-ITA and XX1-CHN isolates, while presence of Asn7 in the same position is noticed in the sequences from cluster II. However, in contrast to the hypothesis of these authors, it seems that change inside this motif sequence is not associated with the specific cluster since PAM-ITA and XX1-CHN isolates sequences are significantly different from the sequences of ApMV strains classified to cluster I and the Polish hazelnut isolates are also grouped separate from both I and II clusters. Cluster analysis revealed that Asn → Gly change in this position was not associated with the host preference as glycine position in MVCKYCGHT motif was noticed in the CP region of ApMV strains from different hosts including hop, elderberry, apricot, peach, Mahaleb cherry, and hazelnut.

The nucleotide sequences of MP gene of the hazelnut isolates were most similar to the sequence of apple isolate KMi2 from China (acc. no. AM490197). All these isolates formed the cluster distinct from the other apple strains which MP gene sequences were available in EMBL/GenBank database. Interestingly, that sequences of CP and MP genes of the hazelnut isolates showed maximum similarity with the corresponding fragments of different reference strains of ApMV. CP gene sequences of the Polish hazelnut isolates were most closely related to the almond strain PAT-ITA from Italy, while the sequences of MP gene of ApMV from hazelnuts showed maximum identity to the apple isolate KMi2 from China.

There are several reports on variability of the CP gene of ApMV [11, 14, 17, 18, 27] and MP gene [14, 27]. However, for the first time, isolates of ApMV from hazelnut were characterized based on the both, CP and MP genes sequences. This is also the first report on detection and identification of ApMV in hazelnut from Poland using molecular biology techniques.

ACKNOWLEDGMENTS

This work was supported by National Science Centre Grant No. 2011/01/B/NZ9/01750.

REFERENCES

1. Akbaş B, Değirmenci K (2009) Incidence and natural spread of Apple mosaic virus on hazelnut in the West Black Sea coast of Turkey and its effect on yield. J Plant Pathol 91(3):767–771

2. Alrefai RH, Shiel PJ, Domier LL, D'arcy CJ, Berger PH, Korban SS (1994) Nucleotide sequence of Apple mosaic virus coat protein gene: no homology with other Bromoviridaecoat protein genes. J Gen Virol 7:2847–2850

3. Aramburu J, Rovira M (1995) Effects of apple mosaic virus (ApMV) on the growth and yield of 'Negret' hazelnut. Acta Hortic 386:565–568

4. Aramburu J, Rovira M (1998) The effects of apple mosaic ilarvirus (ApMV) on hazelnut (Corylus avellana L). J Hortic Sci Biotechnol 73:97–101

5. Aramburu J, Rovira M (2000) Incidence and natural spread of Apple mosaic ilarvirus in hazel in North-east Spain. Plant Pathol 49:423–427

6. Arli Sokmen M, Kutluk Yilmaz ND, Mennan H, Sevik MA (2005) Natural weed hosts of Apple mosaic virus in hazelnut orchards in Turkey. J Plant Pathol 87(3):239–242

7. Atanasoff D (1935) Old and new virus diseases of trees and shrubs. Phytopathol Zeitsch 8:97–223

8. Boom R, Sol CJA, Salimans MMM, Jansen CL, Wertheim-Van Dillen PME, Van Der Nordaa J (1990) Rapid and simple method for purification of nucleic acids. J Clinic Microbiol 28:495–503

9. Dhingra KL (1972) Transmission of Apple mosaic by natural root grafting. Indian J Hort 29:348–350

10. Gottlieb AR, Berbee JG (1973) Line pattern of birch caused by apple mosaic virus. Phytopathology 63:1470–1477

11. Grimová L, Winkowska L, Ryšánek P, Svoboda P, Petrzik K (2013) Reflects the coat protein variability of apple mosaic virus host preference? Virus Genes 47:119–125

12. Hunter JA, Chamberlain EE, Atkinson JD (1958) Note on the transmission of Apple mosaic by natural root grafting. New Zealand J Agricult Res 1:80–82

13. Kobyłko T, Nowak B, Urban A (2005) Incidence of Apple mosaic virus (ApMV) on hazelnut in South-East Poland. Folia Hortic 17(2):153–161

14. Lakshmi V, Hallan V, Ram R, Ahmed N, Zaidi AA, Varma A (2011) Diversity of Apple mosaic virus isolates in India based on coat protein and movement protein genes. Indian J Virol 22:44–49

15. Lee GP, Ryu KH, Kim HR, Kim CS, Lee DW, Kim JS, Park MH, Noh YM, Choi SH, Han DH, Lee CH (2002) Cloning and phylogenetic characterization of coat protein genes of two isolates of apple mosaic virus from 'Fuji' apple. Plant Pathol J 18:259–265

16. Pethybridge SJ, Wilson CR, Hay FS, Leggett GW, Sherriff LJ (2002) Mechanical transmission of Apple mosaic ilarvirus in Australian hop (Humulus lupulus) gardens. Ann App Biol 141:77–85

17. Petrzik K, Lenz O (2002) Remarkable variability of apple mosaic virus capsid protein gene after nucleotide position 141. Arch Virol 147(7):1275–1285

18. Petrzik K (2005) Capsid protein sequence gen analysis of Apple mosaic virus infecting pear. Eur J Plant Pathol 111:355–360

19. Piskornik Z, Kobyłko T, Nowak B (2002) Detection of apple mosaic virus (ApMV) on hazelnut (Corylus sp.) in Poland. Phytopathol Pol 23:31–37

20. Postman JD, Mehlenbacher SA (1994) Apple mosaic virus in hazelnut germplasm. Acta Hortic 351:601–605

21. Postman JD, Cameron HR (1987) Apple mosaic virus in U.S. filbert germplasm. Plant Dis 71:944–945

22. Ragozzino A (1980) Hazel mosaic. Acta Hort 94:375–378

23. Shiel PJ, Alrefai RH, Domier LL, Korban SS, Berger PH (1995) The complete nucleotide sequence of Apple mosaic virus RNA-3. Arch Virol 103:1247–1256

24. Shiel PJ, Berger PH (2000) The complete nucleotide sequence of apple mosaic virus (ApMV) RNA 1 and RNA 2: ApMV is more closely related to alfalfa mosaic virus than to other ilarviruses. J Gen Virol 81(1):273–278

25. Tamura K, Dudley J, Nei M, Kumar S (2007) MEGA4: Molecular Evolutionary Genetics Analysis. (MEGA) software version 4.0. Mol Biol Evol 24:1596–1599

26. Tzanetakis IE, Martin RR (2005) First report of strawberry as a natural host of Apple mosaic virus. Plant Dis 89:31

27. Valasevich N, Cieślińska M, Kolbanova E (2015) Molecular characterization of Apple mosaic virus isolates from apple and rose. Eur J Plant Pathol 141:839–945

28. Wong SM, Horst RK (1993) Purification and characterization of an isolate of apple mosaic virus from rose in the USA. J Phytopathol 139:33–47

Chapter 2

Management of Soilborne Diseases in Crop Plants: An Overview

V. Divya Rani[1*] and H. Sudini[2]

[1]*Department of Plant Pathology, College of Agriculture, Acharya NG Ranga Agricultural University, Hyderabad, India*

[2]*International Crops Research Institute for the Semi-Arid Tropics (ICRISAT), Patancheru, India*

ABSTRACT

Soilborne diseases are very critical in realizing the yield potential of improved cultivars in several agricultural crops. Often these diseases are very difficult to manage due to their highly heterogeneous incidence and lack of knowledge on the epidemiological aspects of soilborne pathogens. Soilborne diseases of ancient and modern agricultural crops have always had some impact on growth and productivity. Observations and experience passed down through generations gave rise to cultural practices that reduced or minimized the losses to soilborne and above ground diseases, but it was probably rare that anyone in agriculture claimed to "control" the diseases. The expansion in the crop diversity in agriculture has required parallel expansion of strategies to minimize the soilborne diseases. The effective control of the soilborne diseases is possible only through detailed study on survival, dissemination of soilborne pathogens; effect of environmental conditions role of cultural practices and host resistance and susceptibility will play a major role in disease management.

Keywords: Soilborne diseases, Soilborne pathogens, Cultural practices, Chemical control

INTRODUCTION

"The diseases that are caused by pathogens which persist (survive) in the soil matrix and in residues on the soil surface are defined as soilborne diseases".Thus the soil is a reservoir of inoculum of these pathogens, the majority of which are widely distributed in agricultural soils. However, some species show localised distribution patterns. Most often damage to root and crown tissues of the plant is hidden in the soil. Thus these diseases may not be noticed until the above-ground (foliar) parts of the plant are affected severely showing symptoms such as stunting, wilting, chlorosis and death.

Soilborne diseases are difficult to control because they are caused by pathogens which can survive for long periods in the absence of the normal crop host, often have a wide host range including weeds, chemical control often does not work well, is not practical or is too expensive and it is difficult to develop resistant varieties of plants. These diseases are often very difficult to diagnose accurately and the pathogens may be difficult to grow in culture and identify accurately. Because of their microscopic size and non-specific symptoms of an infection, soilborne organisms live out of sight and, generally out of mind of the growers and plant protection workers. The effective control of the soilborne diseases is possible only through detailed study on survival, dissemination of soilborne pathogens; effect of environmental conditions role of cultural practices and host resistance and susceptibility will play a major role in disease management [8].

Biology of soilborne pathogens

Survival: Soilborne pathogens survive assoil inhabitants (survive in soil for relatively longer periods), soil invaders or soil transients (survive in the soil for relatively shorter periods). Soilborne pathogens also survive as non-pathogenic and generally in the form of saprobes (organisms that live on decaying organic matter).Under certain congenial conditions these saprobes will turn into pathogenic form.

Distribution of pathogens in the soil

The horizontal and vertical distribution of soilborne pathogens depends on production practices, cropping history, and a variety of otherfactors. Along a vertical axis, the inoculum of most root pathogens lies within the top 10inches of the soil profile, the layers where host roots and tissues and other organic substrates are found. On the horizontal plane, distribution of inoculum in a field is usuallyaggregated in areas where a susceptible crop has been grown:

Factors that influence the distribution: Many factors in the soil influence the activityof soilborne pathogens and diseases: soil type, texture, pH, moisture, temperature, and nutrient levels are among them. Soils that drain poorly, however, tend to favour the survival and distribution of soilborne pathogens such as Pythium, Phytophthora, and Aphanomyces. Similarly, Fusarium and Verticilliumwilts can also bemore severe in wet soils than in dry soils. Only a few root diseases are favoured bydrier soils (for example, common scab of potato caused by Streptomyces scabies).

PREDOMINANT SOILBORNE PATHOGENS:

Fungi: Sclerotium rolfsii, Rhizoctoniasolani, Fusariumsp, Pythium, Phytophthoraetc.

Bacteria: Erwinia, Raltsonia, Rhizomonas, Agrobacterium, Streptomyces etc.

Nematodes: Meloidogyne, Heterodera, Longidorus, Paratrichodorus etc.

Different types of diseases caused by soilborne plant pathogens

Root rots

Soilborne diseases are caused by a diverse group of fungi and related organisms. The most important genera include Pythium and Phytophthora, Rhizoctonia, Cylindrocladium and Armillaria which causes root rots. These diseases are characterised by a decay of the true root system; some pathogens are generally confined to the juvenile roots whilst others are capable of attacking older parts of the root system. Symptoms that are observable include wilting, leaf death and leaf fall, death of branches and limbs and in severe cases death of the whole plant. Some examples of these diseases are shown below.

Rhizoctonia Root Rot Disease:

- Known as damping-off, wire stem, head rot, crown rot
- In older seedling the invasion of the fungusis limited to the outer corticaltissues which develop elongate tan to reddish – brown lesion.
- The region may increase in length and width until they finally girdle the stem and the plant may die.

Stem, collar and head rots:

These diseases are also caused by a diverse group of pathogens including species of Phytophthora, Sclerotium, Rhizoctonia, Sclerotinia, Fusarium and occasionally Aspergillusniger. The most obvious symptom of these diseases is the decay of the stem at ground level. Quite often this decay can lead to symptoms of wilting, death of leaves and to death of the plant. Some examples of these diseases are shown below.Phytophthora spp. for example, can cause diseases such as heart rot of pineapple, blight of potato and tomato and some fruit rots in these conditions. Similarly Rhizoctonia spp. can cause leaf blight in maize and head rot of cabbage in warm wet weather.

Wilts

The main species of fungi that cause these diseases are Fusariumoxysporum and Verticillium spp. The symptoms of these diseases include wilting of the foliage and internal necrosis of the vascular tissue in the stem of the plant. Some species of bacteria can also cause similar types of diseases. Some examples of these diseases are shown below.

Seedling blights and damping off diseases

Various common names are used for diseases of seedlings such as seedling blight and damping-off. The fungi that commonly cause seedling death include Pythium, Phytophthora, Rhizoctonia, Sclerotium rolfsii and less commonly Fusarium sp. These fungi can infect the seedling during the germination, pre-emergence or postemergence phases of seedling establishment. In northern Vietnam Pythium, Rhizoctonia and Sclerotium rolfsii are commonly associated with seedling death of vegetables such as beans, cabbages and other cruciferous crops, cucurbits and tomato.

Pythium Damping –off Disease

The species most often encountered are Pythiumdebaryanum, P. ultimum, P. aphanidermatum, and P. graminicola.The disease often occurs in a roughly circular pattern. This is because of the tendency of fungi to grow radically from the point of origin, which is one of the field markers to distinguish the diseases between other factors that cause the same symptoms.

Phytophthora damping-off:

- Phytophthoraspecies belong to class Oomycetes, family Pythiaceae.
- P. cactorum, P. fragaria, P. palmivora, and P. syringaecause primarily low stem rot, damping – off of vegetables, forest trees, and ornamentals.
- Unlike Pythium, Phytophthora is aggressive in warmer soil temperatures (15-23° C), but still cool condition. Flooding, along with warm temperature.
- Initially affected tissue develops a soft, waterybrown rot. Within several days, the affected plantparts may dry.

Management of soilborne diseases

environmental conditions that favours the infection. In order for adisease to develop, all three factors must be present. The pathogen (a virulent, infectious agent) must have viable inoculum, such as zoospores, available to infect thehost. The host (a susceptible plant) must be exposed to the pathogen's inoculum, andbe physiologically susceptible to infection. Finally, the environmental conditions must be favourable for the infection of the plant and growth of the pathogen. For example, the soil must be saturated with water for a certain period of time in order for water moulds to develop and infect roots. An understanding of these pathogen-host-environment dynamics will help you devise a disease management strategy.

An effective disease management option must be economical: that is, the valueof the crop saved must exceed the cost of control. For this reason, assessments of disease incidence, disease severity, and potential crop loss are key factors when considering control strategies. The careful, regular monitoring of fields and the thoroughexamination of symptomatic plants are essential steps. The timing of control measures is also critical. Management of a destructive disease such as Phytophthora root rotmay require early implementation of appropriate management measures. Besidesbeing economically sound, a management strategy should also be simple, safe, inexpensive to apply, and sufficiently effective to reduce diseases to acceptable levels. Fewmanagement options possess all of these desirable qualities, however, so it usually isbest to integrate multiple management options (e.g., planting resistant varieties, following beneficial cultural practices, and applying disease-control materials).

Management options

Management of soilborne diseases is often difficult. Since there is no silver bullet to control these important diseases, it is always advisable to follow all available methods in an integrated manner to better manage soilborne diseases. The methods include,

- Cultural Practices
- Biological Control
- Chemical Control
- Host plant resistance

Cultural control

Fertilizer application:

- Application of fertilizers along with irrigation improves the overall plant health and thereby reduces the impact of severity of diseases.
- Application of ammonium bicarbonate reduce the viability of sclerotial bodies of S. rolfsii
- Application ofphosphatic fertilizers also influences the host resistance by increasing the production of phytoalexins.
- Management of Pythium and Phytophthora by application of phosphoric acid.
- Application of gypsum reduces the incidence of Macrophominain groundnut.

Providing good soil drainage and good air circulation among plants

Management of irrigation to minimize water dispersal of soil borne pathogens and monitoring disease incidence by avoid spread to other areas are practices that have no apparent involvement with soil microbes. When diseases occur timely removal of dead or infected plants can reduce the potential for inoculum build up.

- Good soil drainage reduces the number and activity of certain oomycetes pathogens (eg.,Pythium) and nematodes.
- Flooding fields for long periods or dry fallowing may also reduce Fusarium, Sclerotiniasclerotiorum, and nematodes.
- Irrigation also helps to reduce the soilborne disease charcoal rot caused by M. phaseolina.

Crop rotation

Generally soilborne pathogens survive in the soil and plant debris up to several years. Crop rotation will be helpful to control the soilborne inoculum because if the host is not present for particular number of years then the amount of inoculum will be reduced. Satisfactory control through crop rotation is possible with pathogens that are soil invaders, i.e., survive only on living plants or only as long as the host residue persists as a substrate for their saprophytic existence. When the pathogen is a soil inhabitant, however, i.e., produces long-lived spores or can live as a saprophyte for more than 5 or 6 years, crop rotation becomes less effective or impractical. In the latter cases, crop rotation can still reduce populations of the pathogen in the soil (e.g., Verticillium) and appreciable yields from the susceptible crop can be obtained every third or fourth year of the rotation. In some cropping systems the field is tilled and left fallow for a year or part of the year.

During fallow, debris and inoculum are destroyed by microorganisms with little or no replacement. In areas with hot summers, fallow allows greater heating and drying of the soil, which leads to a marked reduction of nematodes and some other pathogens. Other cropping systems utilize herbicides and reduced tillage and fallow (Eco fallow).

Tillage practices

Soil preparation before sowing helps in reducing pathogen population by either burial of inoculum deep into the soil or its drying in the top exposed layers. Deep ploughing of crop residues which harbor the pathogen is more effective in reducing this important source of infection. Sub-soiling prior to planting was found to increase the green pea yields of root rot susceptible and tolerant cultivars planted in the soil infested with F. solanif.sp. pisiand Pythiumultimum [13]. The integrated use of tolerant cultivars and cultural practices which reduce the soil compaction could economically reduce the effects of pea root rot in sandy loam soils.

Soil amendments:

- Application of organic amendments like saw dust, straw, oil cake, etc., will effectively manage the diseases caused by Pythium, Phytophthora, Verticillium,Macrophomina, Phymatotrichumand Aphanomyces

- Beneficial microorganisms increases in soil and helps in suppression of pathogenic microbes. For example, application of lime (2500 Kg/ha) reduces the club root of cabbage by increasing soil pH to 8.5 [14]. Similarly application of sulphur (900 Kg/ha) to soil brings the soil pH to 5.2 and reduces the incidence of common scab of potato cause by Streptomyces scabies [6].

- Application of castor cake and neem leaves helps to reduce the foot rot of wheat [14].

- Incorporation of de-caffeinated waste and water hyacinth against root knot nematode [5].

[15] was conducted a field experiment during 2006–2007 wheat growing season at Baoding, Hubei Province, China, aiming at exploring the influence of different amendment rates of maize straw on winter wheat soilborne diseases induced by Rhizoctoniacerealis, Gaeumannomycesgraminis and Bipolarissorokiniana in field conditions. The results showed that the occurrence of three soilborne diseases tested was significantly different under different amendment rates. During the greening stage and jointing stage, the disease indexes of three soilborne diseases were reduced significantly by treatments at the maize straw amendment rates of 7500 kg ha–1 and 3750 kg ha–1. However, disease indexes of wheat common rot and sharp eyespot increased dramatically when the amendment rate increased to 15,000 kg ha–1. At the amendment rate of 15000 kg h–1 wheat root vigour and SOD activity decreased, and ion infiltration and cell membrane-lipid peroxidation level increased, respectively. In the meantime, higher amounts of bacteria and actinomycetes were recorded in the 7500 kg ha–1 amendment rate treatment, while a higher amount of fungi was recorded in the 15000 kg ha–1 amendment rate treatment.

Plant density

- Plant density influences the micro-climate and spread of the diseases. Close spacing raises atmospheric humidity and favours sporulation by many pathogenic fungi.
- Wider spacing reduces the F.oxysporumf.sp. ciceri.
- Severity of crown rot population in wheat increased due to high plant population.
- High density planting in chillies leads to high incidence of damping off in nurseries.

Weed control

Both host and non-host weeds favour disease development by influencing the factors resulting in more inoculum. This includes shading the soil and serving as organic substrate or food base for infection of the host crop. These also change the crop micro-climate and affect soil microflora. In case of wheat, level of take- all disease was found to be higher in the crop having weed intensity indirectly. Trifluralin, was found to greatly increase the cotton blight (R. solani) in Israel as it increased host susceptibility and enhanced the saprophytic activity of pathogen, but the same weedicide has been shown to decrease club root on brassicas in greenhouse experiments in England. Similarly spraying of pea crop with MCPB or MCPA increased root rot severity and quickened the expression of root rot symptoms which was due to stress of herbicide on the plant.

Soil Solarization

When clear polyethylene is placed over moist soil during sunny summer days, the temperature at the top 5 centimetres of soil may reach as high as 52°C compared to a maximum of 37°C in un-mulched soil. If sunny weather continues for several days or weeks, the increased soil temperature from solar heat, known as solarization, inactivates (kills) many soilborne pathogens such as fungi, nematodes, and bacteria near the soil surface, thereby reducing the inoculum and the potential for disease.

Verticillium wilt, Fusarial wilt will be controlled by soil solarization. Bacterial canker of tomato, caused by Clavibactermichiganense, isalso reduced by this method.

Sub-lethal doses of temperatures due to soil solarization also make the pathogen propagules more susceptible to attack of biocontrol agents.

Growing of cover crops:

- Mustard and Brassicasp (Broccoli) helps to reduce the load of soilborne pathogens.
- Cover crops will increase soil microbial diversity by enhancing the soil microflora.
- Create unfavorable conditions.
- The Brassica spp with high content of glucosinolates in their tissues and allylisothiocyanate in leaf extracts.
- Effect of the incorporation of leaves and /or roots of B. Junceaagainst Rhizoctoniasolani and Gaeumannomycesgraminisvar. tritici and suggested the existence of a complex interaction between these two types of residues.

Biological control

Biological controlofpathogens, i.e., the total or partial destruction of pathogen populations by other organisms, occurs routinely in nature. There are, for example, several diseases in which the pathogen cannot develop in certain areas either because the soil, called suppressive soil, contains microorganisms antagonistic to the pathogen or because the plant that is attacked by apathogen has also been inoculated naturally with antagonistic microorganisms before or after the pathogen attack. Sometimes, the antagonistic microorganisms may consist of avirulent strains of the same pathogen that destroy or inhibit the development of the pathogen, as happens in hypovirulence and cross protection. In some cases, even higher plants reduce the amount of inoculum either by trapping available pathogens (trap plants) or by releasing into the soil substances toxic to the pathogen. Researchers have increased their efforts to take advantage of such natural biological antagonisms and to develop strategies by which biological control can be used effectively against several plant diseases.

Mechanisms of biological control of plant pathogens

Antibiosis – Inhibition of one organism by another as a result of diffusion of an antibiotic

- Antibiotic production is common in soil-dwelling bacteria and fungi. For example, zwittermicin A production by B. cereus against Phytophthora root rot in alfalfa

Nutrient competition – competition between microorganisms for carbon, nitrogen, O2, iron, and other nutrients

Destructive mycoparasitism – the parasitism of one fungus by another. Direct contact / Cell wall degrading enzymes/Some produce antibiotics.

Example

- Trichoderma harzianum, BioTrek, used as seed treatment against pathogenic fungus

Requirements of successful biocontrol:

1. Highly effective biocontrol strain must

a. Be able to compete and persist

b. Be able to colonize and proliferate

c. Be non-pathogenic to host plant and environment

2. Inexpensive production and formulation of agent must be developed

a. Production must result in biomass with excellent shelf life

b. To be successful as agricultural agent must be

1. Inexpensive

2. Able to produce in large quantities

3. Maintain viability

3. Delivery and application must permit full expression of the agent

a. Must ensure agents will grow and achieve their purpose

4. Trichoderma spp. are present in nearly all agricultural soils

5. Antifungal abilities have been known since 1930s

1. Mycoparasitism

2. Nutrient competition

6. Agriculturally used as biocontrol agent and as a plant growth promoter

Suppressive Soils

Several soilborne pathogens, such as Fusariumoxysporum (the cause of vascular wilts), Gaeumannomycesgraminis (the cause of take-all of wheat), Phytophthoracinnamomi(the cause of root rots of many fruit andforest trees), Pythiumspp. (a cause of damping-off), and Heteroderaavenae(the oat cyst nematode), develop well and cause severe diseases in some soils, known as conducivesoils, whereas they develop much less and cause much milder diseases in other soils, known as suppressivesoils. The mechanisms by which soils are suppressive to different pathogens are not always clear but may involve biotic and/or abiotic factors and mayvary with the pathogen.

In most cases, however, it appears that they operate primarily by the presence in such soils of one or several microorganisms antagonistic to the pathogen. Such antagonists, through the antibiotics they produce, through lytic enzymes, through competition for food, or through direct parasitizing of the pathogen, do not allow the pathogen to reach high enough populations to cause severe disease. Numerous kinds of antagonistic microorganisms have been found to increase in suppressive soils; most commonly, however, pathogen and disease suppression has been shown to be caused by fungi, such as Trichoderma, Penicillium, and Sporidesmium, or by bacteria of the genera Pseudomonas, Bacillus, and Streptomyces. Suppressive soil added to conducive soil can reduce the amount of disease by introducing microorganisms antagonistic to the pathogen. For example, soil amended with soil containing a strain of a Streptomyces species antagonistic to Streptomyces scabies, the cause of potato scab, resulted in potato tubers significantly free frompotato scab. Suppressive, virgin soil has beenused, for example, to control Phytophthoraroot rot ofpapaya by planting papaya seedlings in suppressive soilplaced in holes in the orchard soil, which was infestedwith the root rot Phytophthorapalmivora.

Mycorrhizal Fungi and Disease Suppression

Among the most beneficial root-inhabiting organisms, mycorrhizal fungi can cover plant roots, forming what is known as a fungal mat. The mycorrhizal fungi protect plant roots from diseases in several ways.

known as a fungal mat. The mycorrhizal fungi protect plant roots from diseases in several ways. • By providing a physical barrier to the invading pathogen. Physical protection is more likely to exclude soil insects and nematodes than bacteria or fungi. However, some studies have shown that nematodes can penetrate the fungal mat.

- By providing antagonistic chemicals. Mycorrhizal fungi can produce a variety of antibiotics and other toxins that act against pathogenic organisms.

- By competing with the pathogen. By increasing the nutrient-uptake ability of plant roots. For example, improved phosphorus uptake in the host plant has commonly been associated with mychorrhizal fungi when plants are not deprived of nutrients, they are better able to tolerate or resist disease-causing organisms.

- By changing the amount and type of plant root exudates. Pathogens dependent on certain exudates will be at a disadvantage as the exudates change.

Two pot experiments were carried out in the greenhouse at Agriculture Research Center, Giza, Egypt, to study the effect of VAM and T. virideas deterrents against R. solani and F. solanidisease on growth and quality of sugar beet [1]. Among all the treatments plants treated with mycorrhiza and T. viriderecorded higher survival rates in sugar beet plants.

Chemical Control

Chemical pesticides are generally used to protect plant surfaces from infection or to eradicate a pathogen that has already infected a plant. A few chemical treatments, however, are aimed at eradicating or greatly reducing the inoculum before it comes in contact with the plant. They include soil treatments (such as fumigation), disinfestation of warehouses, sanitation of handling equipment, and control of insect vectors of pathogens.

Soil Treatment with Chemicals

Certain fungicides are applied to the soil as dusts, liquid drenches, or granules to control damping-off, seedling blights, crown and root rots, and other diseases. In fields where irrigation is possible, the fungicide is sometimes applied with the irrigation water, particularly in sprinkler irrigation. Fungicides used for soil treatments include metalaxyl, diazoben, pentachloronitrobenzene(PCNB), captan, and chloroneb, although the last two are used primarily as seed treatments. Most soil treatments, however, are aimed at controlling nematodes, and the materials used are volatile gases or produce volatile gases (fumigants) that penetrate the soil throughout (fumigate). Some nematicides, however, are not volatile but, instead, dissolve in soil water and are then distributed through the soil.

- Chemicals in plant disease are used to create the toxic barrier between the host surface and pathogen.
- These are applied in the soil as pre and post plant applications. Generally these treatments are being given in high value cash crops.
- Applied as soil fumigation, soil drenching and seed treatment.
- Fungicides like prothiocarb, propamocarb and metalaxyl are useful to control the Oomycetes pathogens.
- Fosetyl – Al is the fungicide which controls the soilborne pathogens when it is used as foliar spray.

[10] conducted an experiment to know the efficacy of fungicides as seed treatment. All the fungicides significantly increased seed germination and plant size and reduced seedling mortality and root infection by F. solaniin bottle gourd, bitter gourd and cucumber.

Best germination was obtained where infested seed of bottle gourd were treated with Aliette (92%) followed by Benlate (90%), Carbendazim, Ridomil, Mancozeb and Vitavax significantly increased germination by 84-88% as compared to control (78%). Effects of fungicides on plant size were significant but varied. Maximum reduction in seedling mortality was obtained where seeds were treated with Topsin-M (4%) followed by RidomilAliette, Benlate, Carbendazim, Mancozeb and Vitavax. Similarly root infection was significantly controlled by fungicidal treatments but with varied effect. Carbendazim and Topsin-M controlled maximum root infection by 6 and 8% respectively.

Host Plant Resistance

Growing of resistance plants is one of the most effective and economical method. Host plant resistance not only reduces the crop losses but lessens the expenditure incurred on disease control as well as reduces the pollution hazards.

- Resistance is of two types:

i) Monogenic (Vertical)

It is also known as race specific or major gene resistance. It is complete and is stable for pathogens having a few pathotypes but breakdown easily in others. In case of cabbage yellows (F. oxysporumf.sp. conglutinans) monogenic resistance is permanent in nature.

ii) Polygenic (Horizontal):

Also known as race non-specific or quantitative resistance. Polygenic resistance is less effective but generally lasts longer.

Host resistance is most effective when combined with cultural and chemical methods.

Transgenic Approaches

Modern DNA technology has made it possible to engineer transgenic plants that are transformed with genes for tolerance of adverse environmental factors, for resistance against specific diseases, or with genes coding for enzymes such as chitinases and glucanases directed against certain groups of pathogens, such as fungi, viruses, and bacteria, or with nucleic acid sequences that lead to gene silencing of pathogens.

Resistance conferred through specific plant genes

There are numerous crops in which plant genes for specific pathogens have been isolated from resistant plants, transferred into susceptible plants, and expressed in these plants. Provided that all the necessary supportinggenes are also transferred and expressed in the new host, some of the formerly susceptible plants now behave asresistant ones. Such resistant plants are subsequently cloned and multiplied, each producing a distinctive line or variety of plant that is resistant to the specific pathogen. When the resistance gene DRR206 from pea was transferred into canola, the transgenic canola plants exhibited resistance to blackleg disease, caused by the fungus Leptosphaeriamaculans, decreased seedling mortality caused by the root pathogen Rhizoctoniasolani, and resulted in smaller leaf lesions caused by Sclerotiniasclerotiorum.

Transgenic Plants Transformed with Genes Codingfor Anti-pathogen Compounds

Genes coding for several pathogenesis-related (PR) proteins, such as chitinases and some glucanases, have been isolated, cloned, and expressed in plants, thereby interfering with the development of certain groups of pathogens and providing resistance to affected plants. Examples of plants transformed with genes coding for anti-pathogenic compounds include peanut plants transformed with antifungal genes that reduced the incidence of Sclerotinia blight, caused by Sclerotinia minorsignificantly compared to susceptible non-transgenic plants [2], [9].

Management through Remote Sensing

Remote sensing is a science/art that permits us to obtain information about an object/a phenomenon through analysis of data obtained through sensory devices without being in physical contact with that object.

Objectives of remote sensing in Plant Pathology

1. Assessment of disease over a vast area
2. To know the relationship of diseases and environment
3. To know the origin and development of epidemics
4. Quantitative assessment of the disease

Remote sensing techniques of importance to Plant Pathology

1. Aerial photography and 2. Satellite imaging

1. Aerial photography: Aerial photography can detect objects on land over a larger area. Colwell(1956) first used remote sensing technique for monitoring stem rust of wheat. He showed that panchromatic colour and especially infrared aerial photography could be used to detect rusts and viral diseases of small grains and certain diseases of citrus. Later, infrared photography was used in England for late blight of potato. The key to distinguish diseased and healthy parts of a crop is to use appropriate film or filter combinations.

The main film types used are panchromatic, infrared, normal colour and colour infrared. The infrared films are preferred because of their superior sensitivity to visible light and to near infrared wavelengths of radiation (700- 900 mμ). The colour infrared or Ektachrome Aero Infrared (Camouflage Detection Film) is superior as it can show the difference between diseased and healthy patches of plants in colour. The healthy foliage is highly reflective to the infrared wavelengths and appears red on this film whereas blighted or diseased foliage has low infrared reflectance and does not appear red in the photograph.

2. Satellite Imaging

Weather satellites

Often cyclones create heavy clouds with rains and an anti-cyclone creates a cloudless sky. All these can be effectively monitored by weather satellites. Sequential pictures show the movement of these systems before they arrive in an area. Therefore by monitoring epidemic favouring systems using a satellite, the disease occurrence on the field can be monitored. Ex: The spread and deposition of stem rust pathogen of wheat is influenced by definite synoptic weather conditions called Indian stem rust rules.

Earth resources technology satellites (LANDSAT, 1972, USA)

LANDSAT covers the entire globe every 18 days scanning the same area at a fixed time. The scanned data is compared for any major differences happened within 18 days. Nagarajan utilized LANDSAT infrared spectral bands 6 (0.7-0.8µm) and 7 (0.8-1.1µm) to differentiate healthy wheat crop of India and severe yellow rust affected crop of Pakistan.

Examples: Coconut root rot and wilt, black stem rust of wheat, citrus canker

Advantages of remote sensing

1. Reveals pattern of disease incidence, intensity and development over large area

2. Data generated by remote sensing is amenable to multidisciplinary approach

3. Gives synoptic view of large areas

4. Data generated is on a permanent scale and is unbiased

5. Data acquisition is fast compared to traditional methods and data analysed is effectively utilized.

6. Satellite data (ERTS) obtains information of an area periodically so that the information can be updated

7. It frequently poses questions for ground investigators which cannot be generated by ground parties

In the United States, USDA-ARS employed remote sensing technology for detecting crop disease and assessing its impact on productivity include using CIR photography to identify circular areas affected by cotton root rot, Phymatotrichumomnivorum[7] and to estimate yield losses caused by blackroot disease in sugar beets [12]. [3] also demonstrated the potential for aerial video imagery to detect P. omnivoruminkenaf, a crop whose tall growth habit makes it almost impossible to survey from the ground. The TIR can provide early, sometimes previsual, detection of diseases that interfere with the flow of water from the soil through the plant to the atmosphere. As an example, [11] found that cotton plants whose roots were infected with the soilborne fungus P. omnivorum and sugar beets infected with Pythiumaphanidermatum both displayed sunlit leaf temperatures that were 3 to 5°C warmer than adjacent healthy plants. Much more research is required when using remote sensing for identifying specific diseases or when separating them from other causes of plant stress. Hyper spectral techniques are likely to provide some assistance, but coupling existing techniques with weather driven computer models of disease development will probably provide the best approach.

CONCLUSION

Management of soilborne diseases is most successful and economical when all the required information pertaining to the crop, disease affecting it, history of these in the previous years, resistant levels of the host and environmental conditions to prevail is available. Combination of disease management practices may have additive or synergistic effects and such an approach is especially desirable in the case of soilborne diseases which are entirely different epidemiologically. Hopefully, the present situation, which emphasizes the use of integrated disease management practices, will stimulate the development of non-chemical methods of disease management to better manage the soilborne pathogens.

General preventive measures to restrict soilborne diseases occurrence

- Select resistant plant cultivars and certified disease-free stock, then plant them at the right time of year
- Avoid mechanical damage to plants
- When possible, obtain plants in their bare root form; if this isn't possible, remove the potting soil from the root system (removing the unknown soil also gives you the opportunity to inspect the root system)
- Do not over-fertilize; doing so inhibits proper root development
- Management of soil pH—raising the soil pH to 6.5–7 by using nitrate nitrogen in place of ammonical nitrogen, for example will decrease the development of Fusarial Wilt.
- Avoid contamination of the growing medium and purchase quality seed

- Avoid use of low, poorly drained areas for nursery production
- Restrict use of machinery in infested areas, particularly when soil is wet
- Prevent movement of soil from infested to non-infested areas of nursery
- Avoid over watering to puddle or run-off point
- Avoid movement of infected trees within and between nurseries

REFERENCES

1. Aly, M.H & Hussein Manal, Y.2009Vesicular-ArbuscularMycorrhiza and Trichoderma virideas deterrents against soil-borne root rot disease of sugar beet. Sugar Technology. 11(4), 387-391.

2. Chenault, K.D., Payton, M.E &Melouk, H.A.2003. Greenhouse Testing of Transgenic Peanut for Resistance to Sclerotinia minor. Peanut Science, Vol. 30, No. 2, pp. 116-120.

3. Cook, C.G., Escobar, D.E., Everitt, J.H., Cavazos, I. Robinson, A.F.,& Davis M.R.1999Utilizing airborne video imagery in kenaf management and production.Industrial Crops and Products.9: 205–210.

4. Colwell, R. N. (1956) Determining the prevalence of certain cereal crop diseases by means of aerial photography. Hilgardia, 26, 223-286

5. Davis, G & Das, P.1998.Effect of different organic amendments for the management of root-knot nematode, Meloidogyne

Incognita on carrot. Indian Journal of Nematology.28(2):203-207.

6. Davis, J.R., Garner, J.G &Callihan, R.H. 1974. Effects of gypsum, sulfur, terraclor and terraclorsuper-x for potato scab control. American Potato Journal. 51(2): 35-43.

7. Heald, C.M.,Thames, W.J.&Wiegand, C.L.1972. Detection of Rotylenchulusreniformis infestations by aerial infrared photography.Journal of Nematology, 4,299–300.

8. Jones, L.R. 1926. The relation of fungi to soil deterioration. Journal of American Society of Agronomy.18,150-153.

9. Livingstone, D.M., Hampton, J.L., Phipps, P.M., &Grabau E.A. 2005.Enhancing Resistance to Sclerotinia minor in Peanut by Expressing a Barley Oxalate Oxidase Gene.Plant Physiology. 137(4): 1354-1362.

10. Nasreen Sultana &Ghaffar, A. 2010.Effect of fungicides, microbial antagonists and Oilcakes in the control of Fusariumsolani, the cause of seed rot, seedling and root infection of bottle gourd, bitter gourd and cucumber.Pakisthan Journal of Botany. 42(4), 2921-2934.

11. Pinter, Jr., P.J., M.E. Stanghellini, R.J. Reginato, S.B. Idso, A.D. Jenkins, and R.D. Jackson, 1979. Remote detection of biological stresses in plants with infrared thermometry, Science, 205(4406):585–586.

12. Schneider, C.L., &Safir, G.R.1975. Infrared aerial photography estimation of yield potential in sugarbeets exposed to blackroot disease. Plant Disease Reporter.59(8), 627–631.

13. Singh, R.S. 2001. Disease management- The Practices. Introduction to Principles of Plant Pathology.Oxford& IBH Publishing Co. Pvt. Ltd. New Delhi. pp. 310

14. Utkhede, R.S and Guptha, V.K. 1996.Management of Soil Borne Diseases.Kalyani Publishers. New Delhi.

15. Wenchao Zhen., Shutong Wang., Chengyin Zhang., &Zhiying, M.A.2009.Influence of maize straw amendment on soil-borne diseases of winter wheat. Frontline Agriculture China. 3(1), 7–12.

Chapter 3

Tubercle Disease (*Xanthomonas Beticola*) and other Gall-Malformed Diseases of Sugar Beet Roots: A Review

Ewa Moliszewska[1] • Małgorzata Nabrdalik[1] • Jacek Piszczek[2]

[1]*Department of Technical and Environmental Sciences Opole University Opole Poland*

[2]*Regional Experimental StationInstitute of Plant Protection – National Research Institute Toruń Poland*

ABSTRACT

The sugar beet (*Beta vulgaris*) is an important plant in agriculture and sugar industry, and it is widely cultivated in European countries. Getting proper raw material of sugar beets (roots) is a problem for agriculture. Some disease symptoms observed on sugar beet roots are atypical tumor-like deformations. The causative agent of these deformations is known in the old literature as *Xanthomonas beticola.* The disease's name in Poland is "tuberkuloza" and in the USA it refers to a description of a pocket disease—therefore we may consider those diseases to be the same. The clear description of *X. beticola* disease can be found in many phytopathological manuals printed in the past and nowadays. Symptoms of the disease were noted in Poland last year, and the preliminary data of the yield quality show that the quality of diseased roots is worse (less sugar content) than of healthy roots. For the proper disease diagnoses, the literature was searched and this searching lead us to conclusion that there is no simple way to recognize the causal organism in the field conditions, and we suppose that *X. beticola* does not exist.

Keywords: Sugar beet Gall *Xanthomonas beticola*

INTRODUCTION

The sugar beet (*Beta vulgaris*) is an important plant in the agriculture and in the sugar industry, and it is widely cultivated in European countries. When the first sugar factory was constructed and established at Cunern in Lower Silesia in 1801, cultivation of beet for processing started in the same year. Judging according to the current criteria, the first yield of sugar beet roots was not satisfactory, but it was the first step to development of sugar industry. Since that time sugar beet growers have been trying to obtain better and better quality of roots. This means higher sugar content and lower concentrations of amino nitrogen and sodium in cells of a root. At the same time, the aim was to increase the mass of roots obtained from a hectare of a sugar beet plantation; however, there are several possibilities to decrease the yield. The most important ones are diseases which can develop on roots and cause damage of tissues or lower either the sugar content or the sugar yield. Development of the sugar industry has led to obtain of a phytopathological knowledge, which resulted in the descriptions of sugar beet diseases and explaining their importance for the agriculture. The disease symptoms observed on sugar beet roots are unusual tumor-like deformations. The causative agent of these deformations is known in the old literature as *Xanthomonas beticola*. The disease symptoms take the form of multiple nodules grown on the upper surface of the roots. In extreme cases, roots with a large number of tumors are strongly deformed. Occasionally the disease can be confused with the tuberosity of the roots caused by *Rhizobium radiobacter* (syn.*Agrobacterium tumefaciens*).

Finding and detecting the tubercle disease and other gall-malformed diseases of sugar beet roots

Field observations

In 2014 and 2015 we found sugar beet roots with severe gall symptoms. Roots with malformations were found for the first time in 2014 in Kłodawa, and they occurred rather occasionally (Fig. 1). In 2015 symptoms were noted in three localizations—again in Kłodawa (Fig. 2) in the same localization as in the previous year and in two localizations near Toruń (Kłodawa and Toruń are located in central Poland) (Fig. 3). These galls were irregular and developed on the whole root or on the upper part of it. If the whole roots were malformed, then the galls were bigger (Figs. 1, 2) than in the case of roots with symptoms on the upper part (root crown). If the symptoms developed only on the root crown, the galls observed were numerous and tiny (Fig. 3). The preliminary data of the yield quality showed that the quality of diseased roots was worse (less sugar content) than of the healthy roots (Table 1).

Table 1. Yield quality of sugar beet roots with tubercle disease symptoms

Type of roots	Field localization	Sugar content (%)	K (mmol/kg)	Na	N-α	The theoretical sugar yield by Reinefeld (%)	The theoretical sugar yield by Buchholz (%)	Sugar content reduction [Reinefeld/Buchholz] (%)
Healthy	Toruń	16.49	50.5	5.7	21.3	14.07	14.22	-
Diseased		15.47	63.2	7.8	23.2	12.53	12.98	1.54/1.24
Healthy	Kłodawa	17.95	52.8	3	27.2	15.49	15.55	-
Diseased		17.09	57.2	5.7	28	14.38	14.58	1.11/0.97

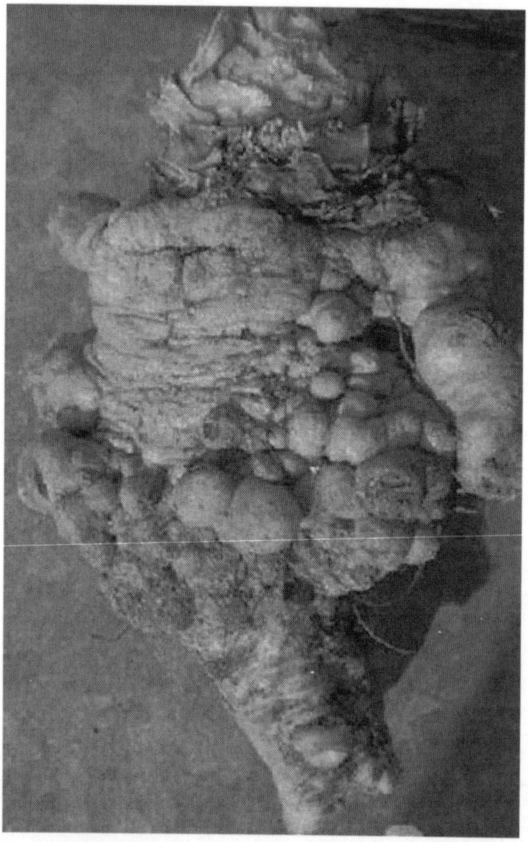

Figure 1. Galls on sugar beet root observed in 2014 (Kłodawa, Poland)

Figure 2. Big galls observed on sugar beet roots, *right*—cross cut of the gall (Kłodawa, Poland)

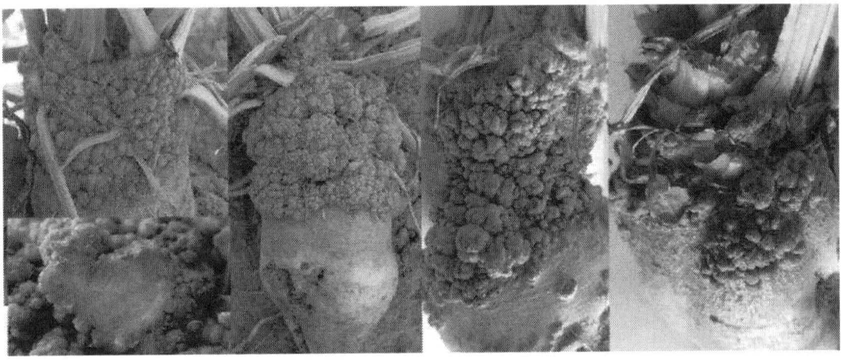

Figure 3. Small galls observed on sugar beet roots, *left down corner*—cross cut of the gall (near Toruń, Poland)

Laboratory investigations

Bacteria present in deformed tissues have been isolated from infested roots. The following growth media specific for *Xanthomonas* type were applied: King B (KB), nutrient agar (NA), nutrient agar supplemented with 0.5 % yeast extract (YNA), nutrient agar supplemented with 1.0 % sucrose (SNA), yeast glucose agar (YGC), sucrose peptone agar (SPA) and malt extract agar (MEA). Incubation was conducted at 28 °C, and observations of colony morphology were made every 24 h for 5 days. The time of colonies' appearance, their size, and morphology were recorded. The isolated bacteria were producing yellow-colored pigment, and their characteristic corresponded with the descriptions given by Benada et al. [2] and Bergey's Manual of Systematic Bacteriology [3]. Preliminary analyses were carried out on the basis of macroscopic and microscopic descriptions as well as biochemical reactions (oxidase, catalase, nitrate reduction, urease, H_2S from peptone, indole, acid production from: glucose, mannose, galactose, cellobiose, lactose, maltose and xylose). Gram-negative bacteria were tested with the use of API 20E and ID 32GN (bioMerieux, France) and the following groups were identified:*Pseudomonas* spp. and *Pantonea* spp. However, none of the isolated bacteria were identified as*Xanthomonas* which urged us to do some research about the disease. The initial laboratory tests did not prove the pathogenic character of bacteria toward beetroot, and thus a detailed study of other research papers was required as well as the search for reference bacteria strains. Further laboratory tests, however, had to be postponed until the following season to collect new infested plants.

Despite the detailed study we carried out, there is still one question to be answered: What is the cause of the malformation/galls on the roots shown in the pictures (Figs. 1, 2, 3)? The symptoms were classified as a tubercle disease of sugar beet (Xanthomonas gall) by experienced researchers and by sugar beet breeders, so we hope that it can be considered a correct diagnose, although the causal agent of it remains unclear.

Current descriptions of Xanthomonas gall disease

Our research on the problem of tubercle disease of sugar beet started from studying the literature, which showed us a great gap of knowledge about this disease. Firstly we classified the symptoms as the tubercle disease of sugar beet or Xanthomonas gall—which we found described in the current and in the old plant disease manuals. We discovered that the disease had not been reviewed and researched for many years, so all of its descriptions seemed to be a copy of the same first information. Pursuit of the information about the disease supposed to be caused by *X. beticola* resulted in the supposition that the bacterium not exists. The results of our investigations, at best, led to the *X.campestris* pv. *betae* or *Pantoea* sp. described from sugar beet. The electronic source of StrainInfo http://www.straininfo.net/taxa/1676 does not contain *X. beticola* [29]. The List of Prokaryotic Names with Standing in Nomenclature (http://www.bacterio.net/) includes the nomenclature of prokaryotes and the nomenclatural changes as cited in the Approved Lists of Bacterial Names or validly published in the International Journal of Systematic Bacteriology (IJSB), or in the

International Journal of Systematic and Evolutionary Microbiology (IJSEM), but there *X. beticola* is not listed. According to Bergey's Manual of Systematic Bacteriology [3]: Volume 2: The Proteobacteria, Part B: The Gammaproteobacteria, *X. beticola* is not a valid name. *X. beticola* was not listed in the first edition of Bergey's Manual of Systematic Bacteriology [13] and in phytobacteriology principles and practice by Janse [11].

Xanthomonas beticola as a cause of pocket disease

The described disease name in Poland is "tuberkuloza" and in the USA it refers to the description of a pocket disease—therefore we may consider those diseases to be the same [6, 10,25]. The clear description of *X. beticola* disease can be found in Benada, Špaček and Šedivy manual of sugar beet pests and diseases printed in Poland in 1984. The bacterium is characterized as aerobic, Gram-negative, flagellate, non-sporulating rod with dimensions of 0.6–0.8 × 1.5–2.0 μm. It can liquefy gelatin, curdle milk and create a yellow colony, reduce nitrates, and create indole on peptone broth with gelatin. It can also use number of sugars as the source of nourishment and produce acids. The optimal growth temperature is 29 °C, maximum—39 °C and minimum—1.5 °C. Optimal pH 6.5 but minimal pH is 4.5–4.8 and maximal—9.0–9.5. The bacterium is not resistant to frost but easily can survive in the dry conditions. It can persist in the soil for long time maintaining pathogenic properties for even 14 years. The bacterium infects through wounds and prefers wet soils [2]. The disease is also described in very recent LIZ (Landwirtschaftlicher Informationsdienst Zuckerrübe) manuals (currently printed and online version) [16, 17], listed by APS [25] and by Streets [28], described by Sherf and MacNab [26] and by Lazarev [14]. The photograph of irregular galls attached to the sugar beet crown given by Cooke and Scott [5] is the most popular in the literature, and it is used in LIZ too [16, 17]. Cooke and Scott [5] gave the possibility to distinguish galls formed by *A. tumefaciens* by narrow tissue bridge from those formed by *X. beticola* by a bridge of tissue almost as wide as the gall and by contain internal cavities or pockets within the gall tissue. Lazarev [14] and Sherf and MacNab [26] described *X. beticola*(Smith, Brown, Townsend) Savelescu which causes tuberculosis of the table beet and the sugar beet as well as fodder beets. The disease occurs in the USA and is marked in the territories of the Russian Federation and also in Moldova, Georgia, Armenia, and Ukraine. The morphological and biochemical description of bacterium corresponds to this given by Benada et al. [2]. Additionally Lazarev [14] pointed that high temperature and the relative air humidity (90 % and higher) favor the development of bacterial infections. He also gave information about the yield loss in favorable for the disease conditions. As he suggested, the percentage of diseased roots reaches even 17–21 % in some years (data for Armenia), and in the diseased roots the sugar content is considerably reduced. Infected roots decay quickly under the influence of secondary infection. Control of the disease needs optimal agriculture, including crop rotation, cultivation of relatively resistant varieties, careful removal of plant residues, and treatment of seeds and plants with pesticides [14]. We have also found a note about Galach'yan [8] experiments which were carried out with naturally infected table beets and artificially infected fodder and sugar beets in 1956 in Armenia (USSR). They showed that sugar beet galls caused by*X. beticola*: cf. 38, 554 were seed-transmissible but only to 1.9–7 % under field conditions. The infection also occurred through the soil, especially to injured roots.

How long we know the pocket disease?

Nyvall [22] described bacterial pocket disease in the USA (Colorado, Maryland, Michigan, New Mexico, Utah, Virginia, Wisconsin, and Wyoming). The disease was also called bacterial canker and was caused by *X. beticola* (Smith, Brown and Town) Burkh. He also paid attention to long periods for surviving in soil, entering roots through wounds at or near the crown level. Bacteria were distributed by any means that can move soil, such as irrigation water and machinery. The galls could develop just at the surface of the ground on the crown but might also occur on petioles and lower on the root. The central portion of the galls was water-soaked and yellow due to the presence of the bacteria. As the galls increased in the size, they became rough and fissured. The atypical information for the disease caused by *X. beticola* was that in the later stages of the disease an abnormal number of leaves were developed [22]. It is worth to noting that probably the first data on the tuberculosis of sugar beet occurred in the USA in the beginning of the XX century. Sugar beets collected in 1910–1912 in the USA showed at the crowns definite galls, which were thought to be crown galls (*Agrobacterium* galls), but many of these did not show the typical features of crown gall. Tumors were initiated by nodules and looked like cultivation wounds. Some of the galls were more or less smooth and globose, and looked much like the crown gall. When the galls were cut across there were brown areas inside, which could be traced for some distance in the interior. The causal organism that produced tubercles on sugar beets was named *Bacterium beticolum* in 1911 and the disease was named the bacterial pocket. In 1913 Serbinow described an organism that he called *B. beticola*(according to [4]). The *B. beticola* was a short motile rod, usually paired, but it could occur singly, in clumps, or in chains of 6–10 cells. The parasite presented a yellow color of the young bacterial colonies, which prevailed in most media. The bacterium was Gram-negative, reduced nitrates to nitrites, reduced methylene blue, produced no indole, and slowly liquefied gelatin. The organism was a wound parasite, which stimulated the tubercles to form discolored galls with cavities which usually contained the brown area with fluid. The tubercles with pockets were reproduced on sugar beets and garden beets by wound inoculation. Although in appearance the disease frequently resembled crown gall, it could be easily distinguished from it by cutting through the outgrowth and noting whether or not there were pockets and stained tissue within. In crown gall the tissue typically was white and sound. The disease was known to occur only in soil rich in nitrogenous fertilizers [4].

Traces of *Xanthomonas beticola* in collections and researches

The genus *Xanthomonas* contains phytopathogenic bacteria that are usually yellow pigmented on medium and can cause several diseases of worldwide distributed plants [15, 27]. *X. beticola*is listed in collection of National Collection of Plant Pathogenic Bacteria (NCPPB [21])—there are two strains of the bacteria added to the collection in 1966. The strain *X. beticola* NCPPB No. 1831 was isolated in 1927 from *Beta vulgaris* in the USA by H.A. Elcock. The pathogenicity of bacteria was confirmed prior to initial freeze drying, but the culture's authenticity is not certain. These bacteria are considered probably not xanthomonad and the name is suggested as illegitimate (NCPPB [21]). Our correspondence with Ms. Charlotte Critchley and her verification of the authenticity of the collected strains by means of fatty acid profiling ensured us that both strains NCPPB 1831 and 1927 are viable, and they are *Bacillus pumilus* and not *X. becticola* (according the personal correspondence with Ms. Charlotte Critchley, National Collection of Plant Pathogenic Bacteria, Fera Science Ltd. in April 2016).

The trace in the investigations of the *X. beticola* was the list of *Xanthomonas* spp. in the article of Starr and Stephens [27] where it was clearly pointed out—*X. beticola* (ICPB XB l09Sm^{r}). The bacterium originated from the general stock culture collection of the Department of Bacteriology, University of California at Davis maintained for the International Collection of Phytopathogenic Bacteria (ICPB). The culture stock number leads on straight to the mistake. The number ICPB XB 109 currently belongs to *X. campestris* pv. *betlicola*—the host plant is*Peper betle* [19]. According to personal information, Starr's collection ended up at the University of California at Berkeley, most likely in Plant and Microbial Biology, formerly Plant Pathology (according to correspondence with prof. Richard M. Bostock Dept. of Plant Pathology, University of California and prof. Wolf-Dietrich Heyer University of California, Davis, 2016).

The next trace of *X. beticola* appeared in Ukraine, in 1984. Kozyrovskaya et al. [12] described properties of this pathogen and *Agrobacterium tumefaciens*. It also seems that *X. beticola* was used in test for searching antimicrobial properties of *Bacillus pumilus* by Kolomiets, Romanovskaya and Sverchkova [31]. In recent years nobody has seen pocket disease ("tuberkulez korneplodov" in Russian) in Russia, and old references tell that it was most harmful in Armenia (Ignatov 2016, personal communication).

The other xanthomonad associated with *Beta vulgaris* is *X. campestris* pv. *betae*, and it causes leafspot disease. Other known names for the strain are *X.* pv. *betae* or *X.axonopodis* pv. *betae*(http://www.straininfo.net/strains/23272).

We have found an atypical trace of *X. beticola* in Brazil—Lordello et al. [18] detected *X. beticola* var. *cynarae* n. var. in diseased plants of artichoke (*Cynara scolimus* L.) but no inoculation trial were made as an attempt to clear up the possible pathogenicity of them to*Cynara scolimus*.

Pantoea agglomerans pv. *betae* as a cause of tubercle disease of sugar beet

However, nowadays more often *P. agglomerans* pv. *betae* is considered the cause of this disease in Eastern Europe, especially Ukraine and Russia. Despite the fact that *P. agglomerans* was also isolated from the inner parts of the sugar beet roots with symptoms of rot (Ignatov 2016, personal communication), *P. agglomerans* is also recorded as a necrotizing pathogen of other plants, especially as recognized in Ukraine on soybean plants [23]. The problems which appear in the study of sugar beet pocket disease are due to the wide occurrence in nature of *P. agglomerans* and related species. Additionally it is present almost on all plants and follows other bacterial infections and diseases. As Ignatov stressed *P. agglomerans* contaminates about 25 % of *Xanthomonas* spp. and *Pseudomonas* spp. isolates, but only 10 % of these isolates show hypersensitive response on tested plants and even smaller number have high virulence on host plants. This experience showed him strict dependence of *P. agglomerans* diseases on environment and host reactions, e.g., some pathogenic *P. agglomerans* can occur at temperature above 25 °C and on particular host plant only (Ignatov 2016, personal communicate). The opinion is confirmed by a study of Manulis and Barash [20]. They described*P. agglomerans* pv. *betae* as a parasite of sugar beet. The bacterium was a Gram-negative, non-capsulated, non-spore-forming, predominately motile rod. It was able to form galls at wound sites, mainly in the crown region of the stem. The hosts of *P. agglomerans* pv. *betae* was *Beta vulgaris* and *Gypsophila paniculata*. They described also *P. agglomerans* pv. *gypsophilae*which was pathogenic only on gypsophila but elicited a hypersensitive response on beet. The virulence genes necessary to transform this epiphytic bacterium into the plant pathogen and interactions with the host components have not yet been completed. The biosynthetic pathways for IAA and cytokinin were recognized in *P. agglomerans* pv. *betae*. Inactivation of IAA and cytokinin biosynthesis pathways caused a reduction in gall size but did not eliminate gall initiation. Complete inhibition of galls was achieved by mutations in the *hrp* gene cluster indicating that type III effectors are crucial for gall formation. Observations could suggest that the mode of pathogenic activity was due to the plasmid (pPATH), *hrp* genes cluster, virulence effectors type III and phytohormones produced by both pathogen and host plant, because the growth regulators synthesized by the plant in response to elicitation caused by type III effectors are responsible for gall initiation [7, 20, 30]. The newest data emphasized that crucial for gall formation are phytohormones which are secreted by host plant. The ones produced by pathogen are not absolutely required for gall formation [1]. For diseases control, the use of pathogenic-free transplants and sanitation is suggested [1, 20]. Manulis and Barash [20] and Barash and Manulis-Sasson [1] also pointed out that no resistant cultivars are available, but mobilizing*pthG* (pathogenicity gene on gypsophila) into *P. agglomerans* pv. *betae* (*in trans*) can cause that the beet pathovar will induce a hypersensitive response instead of galls on beet. The function of PthG protein as an Avr protein on beet suggests that this host may possess a resistance gene that recognizes PthG protein. The hypersensitive response was also recognized in other beet species, which suggests that the resistance protein is conserved throughout the genus. The result of economic loss for table beets is that they cannot be processed mechanically [1], but no data concerning the quality of the yield of sugar beets are available.

Tumors on sugar beet roots caused by *Bradyrhizobium betae* sp. nov

Rivas et al. [24] isolated several endophytic slow-growing bacterial strains from tumors of two deformed plants. Phylogenetic analysis of the DNA regions coding 16S rRNA revealed that these strains belonged to the genus *Bradyrhizobium*. Sequence analysis of the 16S–23S rDNA intergenic spacer region indicated that these novel strains formed a homogeneous group different from all *Bradyrhizobium* species previously described. This genus currently includes four species able to produce nodules in several legumes. Rivas et al. [24] suggested, according to phenotypic and molecular taxonomic approaches, that these strains represent a novel species of *Bradyrhizobium* phylogenetically similar to *B. japonicum*. They proposed the name *B. betae*sp. nov. for the novel species of *Bradyrhizobium*.

Tumors on sugar beet roots caused by *Urophlyctis leproides*(*Physoderma leproides*)

Sugar beet crown wart disease develops as an effect of infestation *Beta vulgaris* by *Urophlyctis leproides* (*Physoderma leproides*). Symptoms of the disease are crown and root warts [10]. It was first reported in Algeria in 1894 and since then, the disease has been recorded in Argentina, Italy, Spain, United Kingdom, Palestine and the USA [9]. In Egypt, the disease was observed during harvest in 2003, 2004 and 2005 at three locations in the Nile Delta on different sugar beet cultivars. Disease incidence started from 1 % in 2003 reaching 1–2 % in 2004 and 3 % in 2005. The disease exhibited typical symptoms on root crowns and occasionally on petioles and leaf blades. Affected leaves were malformed showing galls on leaf blades and petioles, galls were greenish brown with a rough appearance. Galls on crowns range the size from 1 cm to 8–10 cm. These galls were spherical, colored green through yellow to brown depending on the age of plant. The surface of tumors was rough and attached to the host by a narrow base. They occurred singly or in complexes. Sections made through a crown gall revealed cavities filled with thick-walled sporangia (resting spores) surrounded by thickened wall. Sporangia were light brown, spherical to ovoid or concave. Resting sporangia are released into the soil when galls are decomposed, and then they could be detected in the soil. Gouda and Emeran [9] demonstrated that Koch's postulates were fulfilled for the pathogen. Symptoms of the disease appeared on the infested plants after 11 days with no symptoms on uninoculated control plants. Finally sporangia were reisolated from the mature galls' tissues [9].

CONCLUSION

Concluding we should point out that there are several possibilities to recognize tumor-like disease symptoms which are typically classified as *Agrobacterium* galls or *Xanthomonas*tubercles. There is no simple way to recognize the proper causal organism in the field condition especially since *P. agglomerans* is involved in the development of gall symptoms on sugar beet roots. Studying the descriptions of tubercle disease/tuberculosis/pocket disease of sugar beet there appears a feeling that various diseases described over the years as one and the same disease were probably caused by various pathogens or they were caused by mixed organisms. We suppose that *X. beticola* does not exist and the description of the disease has been repetitively copied for many years until now. Since we noticed the disease in Poland, in the nearest future a problem seems to arise, especially due to mild winters, and because of it we have to take careful observations of sugar beet fields and continue research programs.

REFERENCES

1. Barash I, Manulis-Sasson S (2009) Recent evolution of bacterial pathogens: the gall-forming *Pantoea agglomerans* case. Annu Rev Phytopathol 47:133–152

2. Benada J, Špaček J, Šedivy J (1984) Atlas chorób i szkodników buraka. PWRiL, Warszawa

3. Brenner DJ, Krieg NR, Staley JT, Garrity GM (2005) Bergey's manual of systematic bacteriology. In: Garrity G, Brenner DJ, Krieg NR, Staley JR (eds) The proteobacteria, part B: the gammaproteobacteria. Springer, East Lansing, p 1106

4. Brown NA (1928) Bacterial pocket disease of the sugar beet. J Agric Res 37(3):155–168

5. Cooke DA, Scott JE (1993) The sugar beet crop. Science to practice. Chapman & Hall, London

6. Drycott AP (2006) Sugar beet. Blackwell Publishing Ltd, Oxford

7. Ezra D, Barash I, Valinsky L, Manulis S (2000) The dual function in virulence and host range restriction of a gene isolated from the pPATH$_{Ehg}$ plasmid of *Erwinia herbicola* pv.*gypsophilae*. Mol Plant Microbe Interact 13(6):683–692

8. Galach'yan RM (1961) The ways of infection of beet gall. In: Panosyan AK (ed) Problems of microbiology. I. [Voprosy mikrobiologii. I.] Mikrobiol. Sborn. Akad. Nauk. Armyan. Akad. Nauk Armyan. S.S.R., S.S.R. Erevan, pp 41–52

9. Gouda MI, Emeran AA (2006) First report of sugar beet crown wart disease caused by *Urophlyctis leproides* in Egypt. New Disease Reports 13:39

10. Harveson RM, Hanson LE, Hein GL (eds) (2009) Compendium of beet diseases and pests, 2nd edn. APS Press, St. Paul

11. Janse JD (2005) Phytobacteriology principles and practice. CABI Publishing, Wallingford

12. Kozyrovskaya NA, Gvozdyak RI, Muras VA, Kordyum VA (1984) Changes in properties of phytopathogenic bacteria effected by plasmid pRD1. Arch Microbiol 137(4):338–343

13. Kreig NR, Holt JG (eds) (1984) Bergey's manual of systematic bacteriology, vol 1. Williams & Wilkins, Baltimore

14. Lazarev AM (2009) *Xanthomonas beticola.* Interactive agricultural ecological atlas of Russia and neighboring countries. Economic plants and their diseases, pests and weeds.http://www.agroatlas.ru/en/content/diseases/Beta_alba/Beta_alba_Xanthomonas_beticola/. Accessed 14 Mar 2016

15. Lee YA, Sung AN, Liu TF, Lee YS (2009) Combination of chromogenic differential medium and *estA*-specific PCR for isolation and detection of phytopathogenic *Xanthomonas* spp. Appl Environ Microbiol 75(21):6831–6838

16. LIZ manuals on-line version (2016) Landwirtschaftlicher Informationsdienst Zuckerrübe.www.int-koop.de/schaeden/mod_liz_schaeden_texttafel/tafel_id/128/partner/bs/lang/en/index.html, http://schaeden.rheinmedia.de/cgibin/schaedlinge_details.cgi?funktion=druckansicht&id=128&sprache=gb&partner=bs. Accessed 14 Mar 2016

17. LIZ—Landwirtschaftlicher Informationsdienst Zuckerrübe (2005) Burak cukrowy choroby i uszkodzenia, szybkie rozpoznanie. Betmor-Wydawnictwo Waldemar Kostrzewa, Poznań**(in Polish: Sugar beet, diseases and injuries, a quick method of identification)**

18. Lordello LGE, Joly S, Campos HR, Camargo L de S (1961) Nematódeos e bactérias em fôlhas de alcachôfra. An Esc Super Agric Luiz de Queiroz 18: 243–249.

19. Mahesha L, Nadugala NS, Amarasinghe BHRR (2009) Diversity among different isolates of*Xanthomonas campestris* pv. *betlicola* on the basis of phenotypic and virulence characteristics. J Natl Sci Found Sri Lanka 37(1):77–80

20. Manulis S, Barash I (2003) *Pantoea agglomerans* pvs. *gypsophilae* and *betae*, recently evolved pathogens? Mol Plant Pathol 4(5):307–314

21. NCPPB—National Collection of Plant Pathogenic Bacteria (2016) *Xanthomonas beticola*.http://ncppb.fera.defra.gov.uk/furtherinfo.cfm?ncppb_no=1927,http://ncppb.fera.defra.gov.uk/furtherinfo.cfm?ncppb_no=1831. Accessed 9 Apr 2016

22. Nyvall RF (1989) Field crop diseases handbook. Springer, Van Nostrand Reinhold, New York

23. Patyka W, Gnatiuk T, Zhytkevych N, Kalinichenko A, Frączek K (2015) Occurrence of the pathogenic bacteria *Pantoea agglomerans* in soybean cultivation. [Występowanie bakterii fitopatogenicznej *Pantoea agglomerans* w uprawie soi]. Prog In Plant Protect 55(3):280–285

24. Rivas R, Willems A, Palomo JL, Garcia-Benavides P, Mateos PF, Martinez-Molina E, Gillis M, Velazquez E (2004) *Bradyrhizobium betae* sp. nov., isolated from roots of *Beta vulgaris*affected by tumour-like deformations. Int J Syst Evol Microbiol 54:1271–1275

25. Ruppel EG (2016) Diseases of beet (*Beta vulgaris* L.), last update 4/8/03, APS—The American Phytopathological Society.http://www.apsnet.org/publications/commonnames/Pages/Beet.aspx. Accessed 9 Apr 2016

26. Sherf AF, MacNab AA (1986) Vegetable diseases and their control, 2nd edn. John Wiley & Sons, New York

27. Starr MP, Stephens WL (1964) Pigmentation and taxonomy of the genus *Xanthomonas*. J Bacteriol 87(2):293–302

28. Streets RB (1984) The diagnosis of plant diseases: a field and laboratory manual emphasizing the most practical methods for rapid identification. Eight printing, The University of Arizona Press, Tucson

29. Verslyppe B, De Smet W, De Baets B, De Vos P, Dawyndt P (2014) StrainInfo introduces electronic passports for microorganisms. Syst Appl Microbiol 37(1):42–50

30. Weinthal DM, Barash I, Panijel M, Valinsky L, Gaba V, Manulis-Sasson S (2007) Distribution and replication of the pathogenicity plasmid pPATH in diverse populations of the gall-forming bacterium *Pantoea agglomerans*. Appl Environ Microbiol 73(23):7552–7561

31. Zaikov GE (ed) (2008) Biotechnology: state of the art and prospects for development. Nova Science Publishers, New York

Chapter 4

Identification of *Phytophthora* Spp. Isolated from Plants and Soil Samples on Strawberry Plantations in Poland

Beata Meszka[1] • Monika Michalecka[1]

Research Institute of Horticulture96-100 SkierniewicePoland

ABSTRACT

Crown and leather rot of strawberry caused by *Phytophthora* spp. are major soil diseases of cultivated strawberry (*Fragaria* × *ananassa*) in Poland. In this study, in total 45 isolates of*Phytophthora* spp. from plants of cultivars Elsanta, Honeoye, Florence, Camarosa, Roxana, Onebor, Alba, Elegance, Albion, Senga Sengana and Malwina originating from crowns (20 isolates) and fruits (25 isolates) as well as 25 soil samples from the same fruiting strawberry plantations located in central, north-eastern, and south-eastern regions of Poland were identified. Among them, 44 isolates from plant organs and eight from soil samples were identified as *Phytophthora cactorum* and from one plant and one soil sample as *P. citricola*; of those remaining from the soil, one was identified as *P. citrophthora* and one as *P. cryptogea*. Identifications were based on morphological characteristics and DNA analyses. Species-specific polymerase chain reaction confirmed the presence of *Phytophthora* spp. in all positive examined plant and soil samples and confirmed that *P. cactorum* is predominantly responsible for crown and leather rot diseases of strawberry in monitored regions of Poland. Pathogenicity tests showed that all isolates of *Phytophthora* spp. originating from soil derived from strawberry plantations were pathogenic both to strawberry and to raspberry.

Keywords: Crown rot *Fragaria* × *ananassa* Identification Leather rot Molecular detection *Phytophthora* spp.

INTRODUCTION

Phytophthora is a genus of plant pathogenic filamentous oomycetes (phylum: Oomycota), which, morphologically, are very similar to the filamentous fungi from the kingdom Fungi [2].*Phytophthora cactorum* was first identified on cacti in 1870 by Lebert and Cohn [12], and it currently infects an extremely large number of plant hosts. It can limit production of many economically important crops, such as apple, pear, rhododendron, azalea, and strawberry [13].*Phytophthora cactorum* can cause root, collar, and crown rots as well as foliar and fruit damage; thus, it is among the most serious soil-borne pathogens affecting crops worldwide [20]. In Poland, *P. cactorum* causes serious strawberry yield losses due to plant stunting and collapse, especially on susceptible cultivars, such as Elsanta, Honeoye, Elegance, Onebor. Stunting of whole plants or wilting of young leaves are the first symptoms and may appear at any time during the season. The pathogen infects the underground parts by motile zoospores and then colonises strawberry crown. As a result, a typical uniform brown discoloration is visible after the crown cross section, which may appear at the base or in the middle of the crown [4]. Severe infection of the crown leads to dysfunction of the vascular system and to dieback of roots. Leaf margins begin to turn brown and entire leaves wilt, often quickly, and, finally, the whole plant dies. Symptoms of strawberry crown rot can be similar to those elicited by*Colletotrichum acutatum* and *Verticillium dahliae*, which are commonly encountered on strawberry plantations in Poland. Therefore, identification of the causal agents of soil-borne diseases on strawberry plantations is an important issue in preventive and pest control management.

The purpose of this study was to investigate (1) which species of *Phytophthora* are causing necrosis on strawberry fruits and plants; (2) which species are occurring in soil samples and are involved in the decline of strawberry plants in central, north-eastern, and south-eastern regions of Poland; and (3) the pathogenicity of *Phytophthora* spp. present in soil for strawberry and raspberry plants.

MATERIALS AND METHODS

Sampling and isolation procedure

In 2013–2014, plants with visible disease symptoms of crown rot and/or leather rot were sampled from 25 strawberry fruiting plantations in the central, north-eastern, and south-eastern regions of Poland (Table 1).

Small fragments of infected plant organs (fruit, root crown) were surface-disinfected by dipping in 0.5 % (w/v) sodium hypochlorite for 2–5 min and rinsed with sterile water before they were placed onto potato dextrose agar (PDA, Sigma-Aldrich Sp.z.o.o, Poznań, Poland) containing streptomycin, in 90-mm Petri dishes. For each plant organ, four Petri dishes were used, each with 10 tissue pieces. Subsequently, plates were incubated at 25 °C until the mycelial growth was observed. After 3–5 days of growth on PDA at 22 °C, isolates were identified by morphological and cultural characteristics [7, 12]. Finally,

Table 1. Strawberry plantations in central, north-eastern, and south-eastern Poland included in this study

Location of strawberry plantation	Cultivar	Area (ha)	Soil texture	Soil pH	Plant material for study
Bobrowiec [C]	Honeoye	0.7	Sandy and loamy clay	6.0	Crown, fruit
Chynów [C]	Onebor	0.5	Sandy clay above clay	6.2	Fruit
Konin [C]	Honeoye	0.3	Podzol and fawn	6.2	Fruit
Godzianów [C]	Honeoye	0.5	Light soil sand	6.2	Crown, fruit
Jankowice [C]	Malwina	0.5	Sandy clay	5.0	Crown, fruit
Skierniewice 1 [C]	Elegance	3.0	Sandy clay	5.5	Crown, fruit
Skierniewice 2 [C]	Elsanta	2.0	Sandy clay	6.4	Fruit
Wilcze Średnie [C]	Florence	2.0	Sandy clay	6.0	Crown, fruit
Teodozjów [C]	Roxana	0.3	Clayey soil	5.5	Fruit
Radziejowice [C]	Elegance	0.7	Sandy clay	6.2	Crown, fruit
Karsznice [C]	Honeoye	2.0	Clayey soil	6.5	Crown, fruit
Zdziarka [C]	Alba	0.5	Sandy clay	5.0	Crown, fruit
Wilga [C]	Malwina	0.7	Sandy clay	6.0	Crown, fruit
Wilków [E]	Onebor	0.5	Sandy clay	5.5	Crown, fruit
Szydłówka [E]	Albion	3.2	Light soil sand	6.0	Crown, fruit
Tuszewo [N–E]	Camarosa	1.0	Sandy clay	5.0	Crown, fruit
Zarzecze [E]	Elegance	0.5	Clayey soil	6.0	Crown, fruit
Huszlew [E]	Elsanta	0.3	Sandy clay	5.8	Crown, fruit
Serniki [E]	Elsanta	0.4	Light soil	6.0	Crown, fruit
Zosinek [E]	Senga Sengana	0.2	Sand	6.0	Crown, fruit
Dzierzkowice-Wola [E]	Honeoye	1.0	Sandy clay	6.2	Crown, fruit
Łubianki [E]	Camarosa	1.5	Clayey soil	6.0	Crown, fruit
Tuszewo [N–E]	Camarosa	1.0	Sandy clay	5.0	Crown, fruit
Orneta [N–E]	Onebor	1.0	Sandy clay	5.5	Fruit
Mrągowo [N–E]	Honeoye	1.0	Loam	5.5	Crown, fruit
Bogumiłowice [S–E]	Honeoye	0.7	Sandy clay	6.0	Crown, fruit

C, E, N, S—region in Poland: *E* eastern, *C* central, *N* northern, and *S* southern

45 isolates of *Phytophthora* spp. from fruits and/or crowns of strawberry cultivars Elsanta, Honeoye, Florence, Camarosa, Roxana, Onebor, Alba, Elegance, Albion, Senga Sengana and Malwina were obtained. The severity of leather rot on strawberry fruits was evaluated on three selected plantations by counting all healthy and infected fruits on 26 plants in each of four randomly chosen plots during harvest. The results were analysed statistically by the variance analysis method on per cent data transformed according to the Bliss function. The differences between means were estimated by the Newman–Keuls test at the 5 % level of significance.

Isolations from soil samples were made from the same 25 fruiting strawberry plantations (Table 1). Bulked soil samples (about 1.0 kg) were collected from 10 randomly chosen positions of each plantation, at a depth of 10–15 cm, and stored at room temperature. For *Phytophthora*spp. isolation, the baiting test according to Themann et al. [23] was conducted. Trays were filled with tap water to about 1 cm above soil level, and leaves of *Rhododendron* cultivar Nova Zembla (8–10 leaves per tray) were floated on the surface. After 5–7 days of incubation at 21–23 °C,*Rhododendron* leaves were harvested, washed under tap and deionised water, and dabbed dry. Then, segments of 5 mm diameter were cut from individual necrotic spots and placed onto PDA in 90-mm Petri dishes, 25 pieces per dish, each in five replicates. Within the next 2 days, hyphae growing from the plated segments were assessed under an optical microscope for the presence of sporangia. Small fragments from the edge of growing microorganisms were transferred onto new PDA plates in order to obtain pure cultures.

Classification of isolates into the *Phytophthora* genus

After 5–7 days of growth on PDA at 22 °C, isolates were identified by morphological and cultural characteristics [12, 18]. The isolates were cultured on PDA in darkness at 25 °C for 14 days. Characteristic morphology of *Phytophthora* spp. mycelial growth and the presence of sporangia and/or oogonia were observed. Additionally, sporangia were produced by flooding round mycelial discs of 6 × 6 mm in diameter taken from growing margins of 3- to 5-day-old colonies on PDA, 1 mm over their surface, with non-sterile soil extract [18] in 90-mm Petri dishes and incubating them in the dark at 18–22 °C. Soil extract was replaced with a fresh one every 24 hours. After three days mycelial fragments were taken from the growing edge of submerged PDA plugs and transferred onto microscopic slides. Zoosporangia and oogonia were observed at 400× magnification under a light microscope Delta Optical Evolution 300 LED (Delta Optical, Mińsk Mazowiecki, Poland).

DNA extraction from *Phytophthora* spp. cultures

From 2- to 5-day-old *Phytophthora* spp. cultures of 45 isolates, about 50–100 mg of fresh mycelia were cut out, and total DNA was extracted following the Aljanabi and Martinez [1] protocol for DNA extraction. DNA was suspended in 50 µl of double-deionised water and kept at −20 °C for further analyses.

PCR with species-specific primers and sequence analysis

The obtained DNA was tested in PCR with the six primer sets (Table 2) specific for the following*Phytophthora* species: *P. citricola*, *P. citrophthora*, *P. cinnamomi*, *P. cryptogea*, *P. cactorum*, and *P. fragariae*. Conditions for PCR were as follows: an initial denaturation step of 3 min at 94 °C, followed by 30 cycles of 30 s of denaturation at 94 °C; 30 s of annealing at 60 °C for *P. fragariae*-specific primers or at 66 °C for the other primers; and 60 s of polymerisation at 72 °C. The final elongation step was 5 min at 72 °C. Each PCR mixture consisted of the following: 10–30 ng of total DNA template, 500 nM of reverse and 500 nM of forward primers of one species-specific set, 0.2 mM of each dNTP, 0.45 unit of DreamTaq Green DNA Polymerase (Thermo Scientific, Vilnius, Lithuania), 1× optimised DreamTaq Green Buffer, and double-distilled water in a total volume of 15 µl. All reactions were performed in a Biometra T3000 Thermocycler (Biometra, Göttingen, Germany). Then, PCR products were separated during electrophoresis on 1.5 % (w/v) gels, stained with ethidium bromide, and visualised using a transilluminator UV (Sigma T1202; Sigma-Aldrich, St. Louis, USA). The obtainment of the proper-size PCR products was confirmed by comparison with the DNA of the reference strains, showing positive signals with the species-specific primers. For three selected isolates, GL1, BT1, and WG1 (identified as *Phytophthora* spp. in the PCR assay), and for those DNA samples derived from *Phytophthora*-like colonies growing in the baiting test for which no PCR product was obtained in species-specific PCRs, the DNA was amplified with ITS1 and ITS4 primers, specific for ITS (internal transcribed spacer) regions of filamentous fungi and oomycetes [26], and subjected to sequence analysis. The obtained DNA sequences were compared with the sequences available in the GenBank database using the Basic Local Alignment Search Tool (www.ncbi.nlm.nih.gov/BLAST/) and deposited in GenBank (National Center for Biotechnology Information), with accession numbers assigned to them.

Table 2. List of primers used in this study and their nucleotide sequences

Primer name	Primer sequence (5′ → 3′)	Target organism	References
Phytophthora spp.-specific primers			
Primer #1	GTCGACGTCCTGCTTGGCACTCTG	*P. citrophthora*	Ersek et al. [11]
Primer #2	CGGTGCTCCGCGACTGTTGTCCAC	*P. citrophthora*	Ersek et al. [11]
CITR1	TCTTGCTTTTTTTGCGAGCC	*P. citricola*	Schubert et al. [21]
CITR2	CGCACCGAGGTGCACACAAA	*P. citricola*	Schubert et al. [21]
ADF1	TACTGTGGGGACGAAAGTCCT	*P. cactorum*	Boersma et al. [5]
ADR1	CCGATTCAAAAGCCAAGCAACT	*P. cactorum*	Boersma et al. [5]
DG9	AACTGAGCTAGTAGCCTCTC	*P. cinnamomi*	Boersma et al. [5]
DC5	CGCCGACTGGCCACACAG	*P. cinnamomi*	Boersma et al. [5]
CRYF2	CGGTTTTCGGCTGGCTGGG	*P. cryptogea*	Boersma et al. [5]
CRYR2	CAGCTTGCGCCAGAACAGAC	*P. cryptogea*	Boersma et al. [5]
DC1	ACTTAGTTGGGGGCCTGTCT	*P. fragariae*	Bonnants et al. [6]
DC5	CGCCGACTGGCCACACAG	*P. fragariae*	Bonnants et al. [6]
Primers for amplification of fungal and oomycetal rDNA region			
ITS1	TCCGTAGGTGAACCTGCGG	Fungi and oomycetes	White et al. [26]
ITS4	TCCTCCGCTTATTGATATGC	Fungi and oomycetes	White et al. [26]

Pathogenicity of *Phytophthora* spp. from strawberry plantations to raspberry and strawberry plants

Pathogenicity tests were conducted with isolates derived from soil samples from strawberry plantations: two isolates of *P. cactorum* (GL1, Serniki, and BT1, Tuszewo) and one of *P.*

citricola(WG1, Serniki). In June 2014, leaf petioles of the strawberry cultivar Elsanta and stems of the raspberry cultivar Polka were taken and washed under tap water, dabbed dry, and transferred to trays onto moist, sterile blotting paper covered with a plastic net. Three-millimetre-diameter discs cut from the growing margins of 5-day-old PDA cultures of the three examined*Phytophthora* spp. isolates were placed on strawberry leaf petioles and stems of raspberry and covered with parafilm. Control plant material received plain PDA discs. Trays were covered with the foil and incubated at 24 °C. For the experiment, each of 10 leaf petioles and canes per isolate–host combination conducted in four replicates was randomised. Length of necrosis was measured on the fifth (raspberry) and eighth days (strawberry) after inoculation. The experiment was repeated twice. In order to fulfil Koch's postulates, the causal agents of observed symptoms were reisolated on PDA plates, identified using molecular tests, and used for reinfection of plants.

Statistical analyses

Results from the different pathogenicity trials were analysed using analysis of variance (ANOVA) followed by the Tukey HSD test for multiple comparisons (STATISTICA 10; StatSoft Polska, Kraków, Poland).

RESULTS AND DISCUSSION

Classification of *Phytophthora* spp. isolates on the basis of morphological traits

On the basis of morphological trait estimation during *Phytophthora* spp. growth in pure cultures (colony characters, conidial and setal morphology, and production of sporangia and/or oogonia), 44 isolates derived from examined plant tissues were classified as *P.*

cactorum and one as a *P. citricola.* Among cultures growing in baiting tests, typical *Phytophthora*-like morphology was observed for 10 soil samples. For fast detection of *Phytophthora* species in strawberry plants, culture-based isolation methods are widely used. On PDA media, *P. cactorum* usually forms a white, loosely matted colony. Sporangia are distinctively papillate and are usually borne terminally. Each sporangium may contain more than 50 zoospores [10]. According to Waterhouse [25], *P. citricola* can be readily differentiated from *P. cactorum* var.*applanata* on the basis of some morphological features. In culture, *P. citricola* tends to grow in a petaloid manner, while *P. cactorum* grows uniformly fluffy. Moreover, the apex of a sporangium of *P. citricola* is flat and wide with shallow thickening, while in *P. cactorum*prominent papillae with deep thickenings can be observed. Described differentiating features were also observed for these species in our study (Fig. 1a, b, e, f). Sporangia and oogonia produced by *P. cactorum* isolates on PDA media were observed after 10 days of cultivation, whereas *P. citricola* growing on PDA produced only oogonia; sporangia of the latter species were observed on the third day of cultivation in non-sterile soil extract. The oogonia of *P. citricola* were similar in size to those of *P. cactorum* (Fig. 1i, h). Our observations of growing cultures obtained in baiting tests were similar to those previously described for *Phytophthora*spp.; therefore, it was also possible to preliminarily classify the oomycetes as *P. citricola* and *P. cactorum* and, additionally, to *P. citrophthora* and *P. cryptogea. Phytophthora citrophthora*showed stellate, rosettal growth on PDA medium (Fig. 1c) and was able to produce sporangia in non-sterile soil extract (Fig. 1g), but oogonia were not observed on PDA medium, nor in soil extract. Colonies of *P. cryptogea* were fairly fluffy and cottony, with lobate margins and a slight floral pattern (Fig. 1d), but they did not produce sporangia nor oogonia on PDA or in soil extract. The lack of the production of oogonia by *P. cryptogea* and *P. citrophthora* is consistent with the previous observations [12, 19] that these species are heterothallic and require crossing with an opposite mating type to form oogonia and oospores.

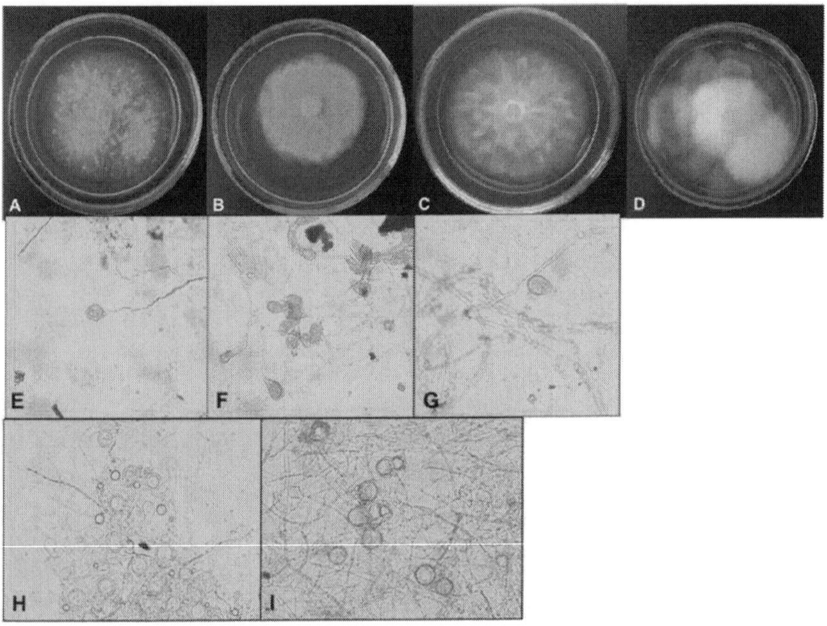

Figure 1. Macroscopic and microscopic features of *Phytophthora* spp. growing in cultures on potato dextrose agar (PDA). The colonies photographed on the sixth day of incubation: **a***P. citricola* (isolate WG1), **b***P. cactorum* (isolate BT1), **c***P. citrophthora* (strain RH04/CTPH), and **d***P. cryptogea* (isolate from Zarzecze). Morphology of zoosporangia of **e***P. cactorum* (isolate BT1), **f***P. citricola* (isolate WG1), and**g***P. citrophthora* (strain RH04/CTPH). Morphology of oogonia of **h***P. cactorum* (isolate BT1) and **i***P. citricola* (isolate WG1). Magnification under light microscope ×400

***Phytophthora* spp. identification in PCR assays**

In PCRs with six *Phytophthora* species-specific primer sets, the presence of the proper-size products of reactions was observed on agarose gel. For 45 tested DNA samples obtained from*Phytophthora* spp. isolates from diseased plant tissues, positive signals for *P. cactorum* were observed (44 isolates) but only one for *P. citricola* (Table 3).

Table 3. Detection of *Phytophthora* spp. in plant and soil samples in the PCR assay

Location of plantation	*P. cactorum*	*P. citricola*	*P. citrophthora*	*P. cryptogea*
Bobrowiec	Crown, fruit			
Bogumiłowice	Crown	Crown		
Chynów	Fruit			
Dzierzkowice-Wola	Crown			
Godzianów	Crown, fruit			
Huszlew	Crown, fruit			
Jankowice	Crown			
Karsznice	Soil, fruit			
Konin	Fruit			
Łubianki	Crown			
Mrągowo	Crown			
Orneta	Fruit			
Radziejowice	Crown			
Serniki	Soil [GL1]	Soil [WG1]		
Skierniewice 1	Soil, crown			
Skierniewice 2	Fruit			
Szydłówka	Soil, crown			
Teodozjów	Fruit			
Tuszewo	Soil [BT1]		Soil	
Wilcze Średnie	Soil, fruit			
Wilga	Crown			
Wilków	Soil, crown			
Zarzecze	Crown			Soil
Zdziarka	Crown			
Zosinek	Soil, crown			

From 25 soil samples, only in 20 % (nine soil samples) were *Phytophthora* spp. strains trapped and recognised in the baiting test. In the PCR assay, isolates from eight plantations were identified as *P. cactorum*, from one as *P. cryptogea* (Zarzecze), from one as *P. citrophthora*(Tuszewo), and from one as *P. citricola* (Serniki) (Table 3). This may suggest that *Phytophthora*spp. propagules are transferred between plantations mostly by strawberry stocks. It has been also reported that *Phytophthora* spp. can survive in infected host tissue [14, 22].

In the case of those *Phytophthora* spp. isolates for which sequencing of the ITS rDNA region and its comparison with reference sequences in GenBank were performed, the species classification was consistent with the results of the PCR assay. BLAST analysis of obtained DNA sequences of GL1 (GenBank Acc. No. KT361202) and BT1 (GenBank Acc. No. KT361201) showed 100 % similarity with the corresponding region of reference of the *P. cactorum* strain (GenBank Acc. No. GU111587.1), while the sequence of the WG1 isolate (GenBank Acc. No. KT361200) was 100 % similar to the sequence of *P. citricola* E strain (GenBank Acc. No. AF266788.1).

For other microorganisms whose colonies were obtained in the baiting test, the sequence analysis of the ITS rDNA region demonstrated the presence of fungi or oomycetes from the genera *Pythium*, *Fusarium*, *Rhizopus*, and *Mortierella*. Species identification based on morphological and physiological characters is a primary approach, although it is mostly laborious and time-consuming. Due to high intra-species variability and small inter-species differences between *Phytophthora* species, additional confirmation is often required. Techniques based on the PCR method offer rapid, simple, and reliable identification of fungal and oomycetal species. Not only can species be identified in pure cultures, but it is also possible to detect pathogens directly in plant material and soil, which highly facilitates the possibility for specific monitoring of these pathogens. On the basis of some species-specific regions in the genome of *Phytophthora*, Ersek et al. [11] developed species-specific primers for *P. parasitica*and *P. citrophthora* detection. The variability in the DNA regions of ITS regions of rRNA of*Phytophthora* spp. has been also examined by Schubert et al. [21] and Boersma et al. [5]; thus, it became possible to discriminate species within the *Phytophthora* genus using species-specific primers designed by those authors. Because of recent changes in the *P. citricola* complex—the new species *P. plurivora* sp. nov. was designated [18]—there was a need to additionally confirm the species affiliation of the WG1 *P. citricola* isolate by comparing its ITS1–5.8S–ITS2 rDNA region with the sequence of reference strains from this complex, available in GenBank database.

Association of *Phytophthora* spp. presence with yield loss

Excessive rainfall during the 2013 growing season was associated with heavy losses in strawberry crop due to fruit rot. Leather rot caused by *P. cactorum* accounted for up to 11 % yield loss depending on cultivar (Fig. 2) within three examined plantations. Propagules of *P. cactorum* are readily dispersed from infected strawberry fruit by splashing water [16]. Gerlach et al. [14] showed that zoospores of *P. citrophthora* were readily dispersed to potted *Pieris japonica* by water. Outbreaks of fruit rot on apple caused by *P. syringae* in the UK were related to the heavy rains, and inoculum was splashed with infested leaf debris on the orchard floor [24]. *Phytophthora* spp. spread mainly through the movement of infested soil, water, and infected plants or plant material; however, there are species that are transmitted aerially [8,15]. Especially substantial as a source of inoculum are those plants that are infected but do not show symptoms either because the disease has not yet progressed to the stage where symptoms are evident. Therefore, the major challenge in preventing the spread of *Phytophthora* is detection of the pathogen in asymptomatic tissue.

Pathogenicity of *Phytophthora* spp. from soil samples to raspberry and strawberry

The average length of necrosis on raspberry cane ranged from 8.0 to 19.7 mm, while on strawberry leaf petioles it ranged from 32.0 to 49.5 mm, depending on the isolate of*Phytophthora* spp. (Table 4). The longest necrosis was observed for isolate WG1 (*P. citricola*), and the smallest, for isolate BT (*P. cactorum*); however, this difference was not significant. All isolates of *Phytophthora* were successfully reisolated from symptomatic tissues and identified using molecular tests, fulfilling Koch's postulates.

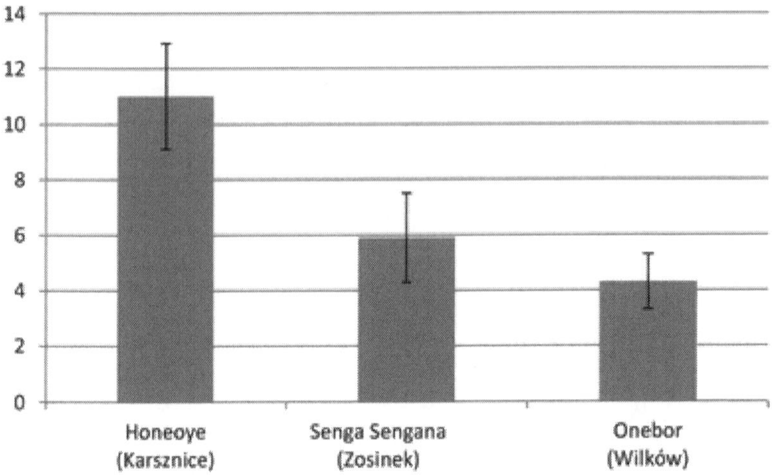

Figure 2. Severity of leather rot of strawberry in the 2013 season

Table 4. Length of necrosis produced on petioles of strawberry and raspberry canes after infection by*Phytophthora* spp.

Name of isolate	Source	Species of *Phytophthora*	Average length of necrosis (mm)	
			Raspberry cane	Strawberry leaf petioles
Control	–	–	0.0 a*	0.0 a
GL1	Soil	*P. cactorum*	14.6 b	42.7 b
BT1	Soil	*P. cactorum*	8.2 ab	32.0 b
WG1	Soil	*P. citricola*	19.7 b	49.5 b
			MS between groups = 119.06	MS between groups = 521.63
			df = 36	*df* = 36

* The same letters indicate lack of statistically significant differences between values in columns, according to the Tukey test

This study shows that two *P. cactorum* and one *P. citricola* strains isolated from soil samples collected in the strawberry plantations where disease symptoms were observed were pathogenic both to strawberry plants and to raspberry plants, causing necrotic lesions on wounded leaf petioles and raspberry canes. The differentiation in *P. cactorum* isolates according to host preference was found in studies of Cooke et al. [9] and Hantula et al. [17]. Bielenin [3] reported that isolates of *P. cactorum* both from strawberry crowns and from fruit caused distinctly smaller necrosis symptoms on apple shoots than isolates from apple trees.

In conclusion, we have shown on the basis of morphology and DNA sequence analysis that the*P. cactorum* species is the most common causal agent of strawberry infections in Poland. Under favourable weather conditions for development and spread of the pathogen (high humidity and temperatures above 20 °C), it causes significant losses in yield of strawberries as well as decay of whole plants. The results of pathogenicity trials revealed that the filamentous parasitic organisms *Phytophthora* spp. (class *Oomycetes*) occurring in the soil can be pathogenic for both strawberry and raspberry plants, showing no host specificity.

ACKNOWLEDGMENTS

The authors would like to thank Magdalena Ptaszek MSc (Department of Vegetables and Ornamental Plant Diseases, Research Institute of Horticulture) and Aleksandra Trzewik MSc (Independent Laboratory of Microscopy, Research Institute of Horticulture) for precious suggestions and remarks concerning *Phytophthora* spp. cultivation and A. Trzewik MSc for providing *P. citrophthora* strain RH04/CTPH.

CONFLICT OF INTEREST

The authors confirm that they have no conflict of interest.

REFERENCES

1. Aljanabi SM, Martinez I (1997) Universal and rapid salt extraction of high-quality genomic DNA for PCR-based techniques. Nucleic Acids Res 25(22):4692–4693

2. Bertier L, Leus L, D'hondt L, de Cock AWAM, Höfte M (2013) Host adaptation and speciation through hybridization and polyploidy in *Phytophthora*. PLoS ONE 8(12):e85385.

3. Bielenin A (2002) Grzyby rodzaju *Phytophthora* w uprawach sadowniczych: występowanie, szkodliwość, I zwalczanie. Zeszyty Naukowe Inst. Sad. i Kwiaciarstwa. Monografie i rozprawy

4. Bielenin A, Meszka B (2009) Choroby krzewów owocowych. Plantpress Sp.z.o.o, Kraków, 129 s

5. Boersma JG, Cooke DEL, Sivasithamparam K (2000) A survey of wildflower farms in the south-west of Western Australia for *Phytophthora* spp. associated with root rots. Aust J Exp Agric 40:1011–1019

6. Bonants P, Hagenaar de Weerdt M, van Gent-Pelzer M, Lacourt I, Dooke DEL, Duncan M (1997) Detection and identification of *Phytophthora fragariae* Hickman by the polymerase chain reaction. Eur J Plant Pathol 103:345–355

7. Brasier CM, Cooke DEL, Duncan JM, Hansen EM (2003) Multiple new phenotypic taxa from trees and riparian ecosystems in *Phytophthora ganapodyides–P. megasperma* ITS Clade 6, which tend to be high-temperature tolerant and either inbreeding or sterile. Mycol Res 107:277–290

8. Cahill DM, Hardham AR (1994) Exploitation of zoospore taxis in the development of a novel dipstick immunoassay for the specific detection of *Phytophthora cinnamomi*. Phytopathology 84:193–200

9. Cooke DEI, Kennedy DM, Guy DC, Russell J, Unkles SE, Duncan JM (1996) Relatedness of Group I species of *Phytophthora* as assessed by randomly amplified polymorphic DNA (RAPDs) and sequences of ribosomal DNA. Mycol Res 100:297–303

10. Ellis MA, Madden LV (1998) Leather rot. In: Maas JL (ed) Compendium of strawberry diseases, 2nd edn. American Phytopathological Society Press, St. Paul, pp 33–35

11. Ersek T, Schoelz JE, English JT (1994) PCR amplification of species-specific DNA sequences can distinguish among *Phytophthora* species. Appl Environ Microb 60(7):2616–2621

12. Erwin DC, Ribeiro OK (1996) *Phytophthora*: diseases worldwide. American Phytopathological Society Press, St. Paul

13. Fujita DB (1990) Crown, collar, and root rot. In: Jones AL, Aldwinkle HS (eds) Compendium of apple and pear diseases. American Phytopathological Society Press, St. Paul

14. Gerlach WWP, Hoitink HAJ, Schmitthenner AF (1976) *Phytophthora citrophthora* on*Pieris japonica*: infection, sporulation, dissemination. Phytopathology 66:302–308

15. Goodwin SB (1997) Population genetics of *Phytophthora*. Phytopathology 87:462–473

16. Grove GG, Madden LV, Ellis MA (1985) Splash dispersal of *Phytophthora cactorum* from infected strawberry fruit. Phytopathology 75(5):611–615

17. Hantula J, Lilja A, Parikka P (1997) Genetic variation and host specificity of *Phytophthora cactorum* isolated from Europe. Mycol Res 101:565–572

18. Jung T, Burgess TI (2009) Re-evaluation of *Phytophthora citricola* isolates from multiple woody hosts in Europe and North America reveals a new species, Phytophthora plurivola sp. nov. Persoonia 22:95–110

19. Jung T, Stukely MJC, St. Hardy GE, White D, Paap T, Burgess TI (2011) Multiple new*Phytophthora* species from ITS Clade 6 associated with natural ecosystems in Australia: evolutionary and ecological implications. Persoonia 26:13–39

20. Maas JL (1998) Compendium of strawberry diseases. 2nd edn. APS Press, St. Paul, p 98

21. Schubert R, Bahnweg G, Nechwatal J, Jung T, Cooke DEL, Duncan JM, Müller-Starck G, Langebartels C, Sandermann H Jr, Oßwald WF (1999) Detection and quantification of*Phytophthora* species which are associated with root rot diseases in European deciduous forests by species-specific polymerase chain reaction. Eur J For Pathol 29:169–188

22. Shea SR, Gillen KJ, Leppard WI (1980) Seasonal variation in population levels of*Phytophthora cinnamomi* Rands in soil in diseased, freely-drained *Eucalyptus marginata*Sm. sites in the northern Jarrah Forest of south-western Australia. Prot Ecol 2:135–156

23. Themann K, Werres S, Luttmann R, Diener HA (2002) Observations of *Phytophthora* spp. in water recirculation systems in commercial hardy ornamental nursery stock. Eur J Plant Pathol 108:337–343

24. Upstone ME (1978) *Phytophthora syringae* fruit of apples. Plant Pathol 27:24–30

25. Waterhouse GM (1957) *Phytophthora citricola* Sawada (syn. *P. cactorum* var. *applanata*Chester). Trans Br Mycol Soc 40:349–357

26. White TJ, Bruns T, Lee S, Taylor J (1990) Amplification and direct sequencing of fungal ribosomal RNA genes for phylogenetics. In: Innis MA, Gelfand DH, Sninsky JJ, White TJ (eds) PCR protocols: a guide to methods and applications. Academic Press, San Diego, pp 315–322

Chapter 5

Efficacy of Selected Plant Extracts Against Pyricularia Grisea, Causal Agent of Rice Blast Disease

Judith Hubert[1,2*], Robert B. Mabagala[1], Delphina P. Mamiro[1]

[1]*Department of Crop Science and Production, Sokoine University of Agriculture, Morogoro, Tanzania*

[2]*Africa Rice Center, Dar es Salaam, Tanzania*

ABSTRACT

Rice blast disease, caused by a seed-borne fungus Pyricularia grisea, is an important and serious disease of rice (Oryza sativa L.) worldwide. The disease has been reported to cause yield losses of up to 40% in Tanzania. Studies were conducted to determine the effect of aqueous extracts of Aloe vera, Allium sativum, Annona muricata, Azadirachta indica, Bidens pilosa, Camellia sinensis, Chrysanthemum coccineum, processed Coffee arabica, Datura stramonium, Nicotiana tabacum and Zingiber officinalis for control of rice blast disease (Pyricularia grisea) in-vitro and in-vivo. The results indicate that processed C. arabica at 10% and 25% (v/v) had the highest (81.12%) and (89.40%) inhibitory effect, respectively, against P. grisea. Aqueous extract from N. tabacum at 10% concentration ranked third (80.35%) in inhibiting P. grisea. These were followed by extracts from

25% A. vera (79.45%) and 25% C. coccineum flower (78.83%). The results also indicate that, extracts from A. indica, A. vera, A. sativum, C. arabica, D. stramonium, C. sinensis, Z. officinalis and N. tabacum did not have any phytotoxic effect on seed germination, shoot height, root length, dry weight, seedling growth and seedling vigour index. These plant extracts can thus be used for rice seed treatment to manage rice blast disease.

Keywords: Plant extracts, Pyricularia grisea, Rice

1. INTRODUCTION

Rice blast disease, caused by a seed-borne fungus Pyricularia grisea, is an important and serious disease of rice (Oryza sativa L.) worldwide [1] - [3] . The disease has been reported to cause yield losses of up to 40% in Tanzania [4] . Rice blast disease is one of the major factors causing losses in quality and quantity of rice [5] . Small- scale farmers in Tanzania keep their harvested rice seeds for sowing in the following season. This practice facilitates multiplication of P. grisea within the seed and becomes a source of inoculum for new infection [6] .

Management of rice blast is difficult because the pathogen is seed-borne. Management approaches have mainly focused on the use of synthetic chemicals and resistant rice varieties [7] . In Tanzania, rice blast management has been limited to foliar spray with fungicides such as Dithane M-45 (80% of Mancozeb) and Benlate (50% of Benomyl) [8] . However, their efficacy can be limited by the rapid development of resistance of the pathogen to these chemicals. Chemicals applied as either seed dressing or spray such as Abendazim, Benomyl, Benomyl + Copper sulphate, probenazole, thiabendazole and pyroquilon fungicides have been used in different rice growing countries to manage rice blast disease [9] . However, such chemicals are expensive, not easily available to small-scale farmers and have detrimental effects on the environment, farmer and consumer health, beneficial predators and parasitoids [10] . These fungicides have also been reported to affect useful organisms used in biological control programs [11] [12] . Amadioha [13] reported water and ethanol leaf extracts and oil extract of seeds of Azadirachta indica to be effective in reducing radial growth of P. grisea in-vitro and development and spread of blast disease in rice plants. Therefore, natural plant products (botanicals) are becoming a new source of agricultural chemicals [13] to manage plant diseases.

Plant extracts have been known for their medicinal and antimicrobial properties since ancient times [14] [15] . They offer a greater scope than synthetic chemicals as they are relatively safe, easily biodegradable and eco- friendly [2] [11] [16] [17] . Gurjar et al. [17] reported that plant extracts from Azadirachta indica, A. Juss., Allium sativum, Linn., Eucalyptus globulus, Labill., Curcuma longa, Linn., Nicotiana tabacum, Linn and Zingibar officinale, Rose inhibited growth of pathogens such as Phytopthora infestans, Alternaria alternate, Rhizoctonia solani and Curvularia lunata. Natural chemicals from plants are cheap, readily available and cost-effective in developing countries where synthetic fungicides are scarce and expensive for resource-poor farmers [7] .

The main focus of the present study was to evaluate efficacy of plant extracts against P. grisea in-vitro and in-vivo and phytotoxicity of plant extracts on rice plant development.

2. MATERIALS AND METHODS

2.1. Seed samples

Rice (Oryza sativa) seed samples of nineteen different rice genotypes were used in this study. Among the collected seed samples, ten (Mbawambili, Masantula, Bishori, Domo la fisi, Kisegese, Mzambia, Kaniki, Karamata, Dongo la Songea and Usiniguse) were local accessions, while Supa, Kihogo red and Afaa Mwanza were varieties commonly grown in many areas. Released varieties were IR 64, Kalalu, Mwangaza,

TXD 85, TXD 88 and Saro 5. The collected seed samples were transported to the African Seed Health Centre (AfSHC) laboratory at Sokoine University of Agriculture and stored at 4˚C for further testing.

2.2. Isolation and Preparation of Pyricularia grisea Inoculum

Cultures of P. grisea used in this study were isolated from infected rice seeds collected from Mvomero, Kilosa and Kilombero districts, Morogoro region, Tanzania. The germinating spores and mycelia were picked with a slightly bent inoculating needle and transferred to the Petri dishes containing Potato Dextrose Agar (31.2 g Potato Dextrose Agar powder and 800 ml of distilled water) at 4 equidistant points [18] . The Petri dishes were sealed with masking tape and incubated at alternating temperatures of about 28˚C - 32˚C for five days in alternating cycles of 12 h light and darkness to induce growth of P. grisea and were kept upside down. The pure cultures of the isolates were grown on V8 agar (3 g Calcium Carbonate, 200 ml V-8 juice and 20 g agar in 800 ml of distilled water) for 14 days to induce sporulation. The V-8 agar was used for producing P. grisea conidia. The petri dishes containing P. grisea inoculum were stored in the refrigerator at 5˚C [19] for further use.

2.3. Sources of plant extracts

Plants used as sources of extracts in this study are listed in Table 1. Fresh leaves (about 3 kg) of each plant species were harvested using a sterilized pair of scissors and put in perforated sacs and labeled properly for location, date and name of plant species. The samples were sent to the African Seed Health Centre (AfSHC) screen house for further processing.

Table 1. Plants used as sources of antifungal extracts in the study.

Common name	Scientific name	Area (region) collected
Neem	*Azadirachta indica*	Morogoro
Aloe	*Aloe vera*	Morogoro
Garlic	*Allium sativum*	Iringa
Black jack	*Bidens pilosa*	Morogoro
Soursop	*Annona muricata*	Morogoro
Coffee	*Coffee arabica*	Morogoro
Thorn apple	*Datura stramonium*	Mbeya
Pyrethrum	*Chrysanthemum coccineum*	Iringa
Tobacco	*Nicotiana tabacum*	Iringa
Ginger	*Zingiber officinalis*	Iringa
Tea	*Camellia sinensis*	Mbeya

2.4. Preparation of plant extracts

Leaves of plants presented in Table 1 were washed with running tape water to remove soil materials, and rinsed with Sterile Distilled Water (SDW) three times. The leaf samples were then cut into small pieces and placed on benches in the screen house at 25˚C - 28˚C to dry for 3 to 4 weeks. The dried leaves of each plant species were made into powder separately using a sterilized mortar and pestle and then sieved with one millimeter sieve. The powder of each plant species was packed in water proof plastic bags and labeled appropriately as described by Akinbode and Ikotun [20] and stored at 4˚C until used [21] . Crude plant extracts were obtained by infusing 50 g of each plant material in 100 ml SDW to give 50% w/v in a 500 ml conical flask and the mixture was incubated at 25˚C - 28˚C for 20 hours [22] [23] . The infusion was filtered separately through sterile double-layered cheese cloth into a sterile 400 ml beaker and the resulting stock solution was collected and stored at 25˚C - 28˚C until used [24] .

2.5. Amendment of PDA with plant extracts

Four different concentrations (1%, 5%, 10% and 25%) of each plant extract were prepared and added directly to the sterile PDA through sterile micro-filters in the laminar flow chamber. The stock solutions were removed and measured by a 10 ml syringe without a needle. To prepare these concentrations, two, 10, 20 and 50 ml of the stock solutions were mixed in 198 ml, 190 ml, 180 ml and 150 ml, respectively, of autoclaved PDA as described by Kamalakannan and Shanmugam [25] . The PDA medium with plant extracts (PDAPE) were poured into Petri dishes (20 ml) and allowed to solidify and then stored at 5°C until used.

2.6. In-Vitro assay of plant extracts against Pyricularia grisea

The fungitoxic effect of different plant extracts in-vitro were studied by inoculating PDA with 14-day-old pure cultures of P. grisea [19] . Small mycelia segments (5 mm) were made using a sterile 5 mm diameter cork borer. Each mycelial segment was then transferred singly onto the centre of each PDAPE Petri dishes at 1%, 5%, 10% and 25% concentrations (v/v). The PDA Petri dishes without plant extracts were used as negative control and PDA mixed with Apron Star® 42 WS contained 20% Thiamethoxam, 20% Metalaxyl-M and 2% Difenoconazole, trade names Cruiser and Actara were included as positive controls. The experiments were laid out in a split plot in a randomized complete block design with rice blast as main plot and plant extracts as a subplots with 50 treatments (12 plant extracts with 4 different concentrations and 2 controls: Apron Star® 42 WS and distilled water). The inoculated Petri dishes were sealed with masking tape and incubated at 28°C - 32°C for five days in alternating cycles of 12 hours light and darkness to induce growth of P. grisea. Radial growth of P. grisea was then recorded after every three days for up to 21 days after inoculation by measuring mycelial growth diameters along two diagonal lines previously drawn on the reverse side of each Petri dish to serve as a reference, using a 30 cm plastic ruler to determine the effectiveness of plant extracts. Calculation of percent inhibition of fungal growth was estimated based on Ogbeborand Adekunle [26] methods as follows: Percentage mycelial inhibition = 100 (Mycelial growth diameter in control—Mycelial growth diameter in treatment)/Mycelial growth diameter in control. The treatments that showed high percentage control in in-vitro by reducing P. grisea growth were selected and the experiments were repeated twice.

2.7. In-Vivo assay of plant extracts against P. grisea

2.7.1. Seed inoculation with P. grisea

Four hundred seeds of a susceptible rice variety Supa were spray-inoculated with 1×10^5 spores/ml of P. grisea strain following procedures of Namai and Ehara [27] . Inoculated seeds were dried in the laminar flow chamber on three layers of blotter papers in Petri dishes for 2 h. Seeds were stored at 4°C until used.

2.7.2. Treatment of inoculated seeds with plant extracts

The most promising concentrations of plant extracts in in-vitro were used in in-vivo. Twenty five per cent suspensions of extracts of Azadirachta indica, Aloe vera, Allium sativum, Coffee arabica, Camellia sinensis, Datura stramonium, Nicotiana tabacum and Zingiber officinalis were prepared by soaking 15 g of plant parts (ground leaves and seeds) in 60 ml of SDW. Ten per cent suspension of Nicotiana tabacum extracts was also prepared by soaking 6 g of leaves in 60 ml of SDW for 20 h [22] . The extracts were filtered using double-layered cheesecloth and the volume was adjusted to 60 ml with SDW and put in 80 ml scotch bottle with cover. The extracts were then kept at 4°C until used.

Rice seeds (pre-inoculated with P. grisea) were soaked in suspensions of each pre-determined concentrations of plant extract overnight and then dried on sterile blotter papers for 2 h on the laminar flow chamber. The seeds were placed in a beaker and 40 ml of each plant extract suspension was added. The seeds were gently stirred by a stirring glass rod to ensure that they were completely immersed and evenly distributed. Beakers containing inoculated seed were covered by aluminum foil to reduce external contaminations and placed at the constant temperature of 25˚C for 20 h.

Two hundred seeds in 4 replicates (50 seeds per replicate) were tested per plant extract. Treated seeds were planted in pots (25 seeds/pot) containing a mixture of sandy-loam soil:farmyard manures (3:1) and kept under screen house conditions (25˚C - 30˚C and 85% - 95% RH). The efficacy of plant extract treatments in the control of P. grisea incidence in rice seedling was obtained by assessing rice seedlings for disease incidence and severity at 15, 20, 25, 30 and 35 days after sowing [28] . Scores for disease severity were done using the international scale (0 - 9 scale) developed by International Rice Research Institute [4] . In addition, the height of seedlings and leaves were evaluated. The weight of seedlings was evaluated 35 days after sowing. The height of the seedlings was determined using a ruler by measuring aerial part of the seedling from the soil surface to the terminal node of the developing leaf. To determine the fresh weight, rice seedlings were carefully cut at the bottom of the stem using a pair of scissors and the seedlings were placed on a weighing balance.

2.8. Evaluation of phytotoxicity of plant extracts on rice seed germination, seedling growth and vigour

The effect of plant extracts on seed germination, seedling growth and vigour tests were carried out to determine whether the plant extracts were harmful. Seed germination tests were carried out using Between Paper method [29] . Four hundred seeds from three different rice varieties (Mwangaza, Saro-5 and Supa) were transferred separately to a beaker containing 40 ml suspensions of extracts from 25% (v/v) of Azadirachta indica, Aloe vera, Allium sativum, Camellia sinensis, Datura stramonium, Nicotiana tabacum, Zingiber officinalis and 10% Coffee arabica based on their best performance in the in-vitro and in-vivo assays. Seeds were soaked in the plant extract(s) for 24 h at 25°C and dried on sterile blotter papers for 2 h on the laminar flow chamber [30] . Rice seeds soaked in SDW and Apron Star® 42 WS treatments were included in the experiments as negative and positive controls, respectively.

The effect of plant extracts on seed germination was evaluated by counting the number of normal seedlings, abnormal seedlings and dead seeds as recommended byInternational Seed Testing Association [29] Seedling Evaluation Handbook. Four hundred seeds in four replicates of hundred seeds of each sample were placed between two layers of moist germination papers. The germination papers were then folded along one edge and rolled up carefully ensuring that no excess pressure was placed on the seeds. The rolled germination papers were tied with rubber bands at both ends and put into the transparent plastic bags tied at the top. To prevent the bending, paper towels were transferred in a basket with a flat base (W × L × H = 9.8 cm × 9.8 cm × 9.8 cm) and then placed in the incubation room at 25°C - 28°C for 14 days. The shoot length, root length and dry weight of seedlings were measured and recorded 14 days after sowing in order to evaluate the effect of the extracts on seedling growth and vigour [29] . The seedling shoots were cut, kept separately in paper bags and dried in an electric oven maintained at 80°C temperature for 24 h as described by Chhetri [31] . After drying, the shoot dry weights were recorded by an electric balance. In order to obtain the seedling growth rate (gram per plant) the total weight of dry seedlings per roll of germination paper towels was divided by total number of normal seedlings as described by International Seed Testing Association [29] Seedling Evaluation Handbook.

2.9. Data analysis

All data collected were analyzed based on the split plot arrangement in a CRD analysis of variance (ANOVA) model using GenStat statistical software. The data for disease incidence and severity of seed-borne fungal pathogens including P. grisea and the level of seed infection per sample and percentage germination test (abnormal seedlings, dead seeds and hard seeds) were subjected to square root transformation to normalize the data before analysis. A constant value (0.5) was added to each observation before taking square root [13] . Means were separated by Duncan's multiple range test (DMRT) at P = 0.05.

3. RESULTS AND DISCUSSION

Results indicate that the concentrations of the tested plant extracts against P. grisea had a positive effect in inhibiting mycelia growth. The results of P. grisea growth on PDA amended with plant extracts showed that A. indica, A. vera, A. sativum, C. arabica, C. coccineum, D. stramonium, C. sinensis, Z. officinalis and N. tabacum had antifungal properties against P. grisea at high (25%) but not at low (1%) concentrations in in-vitro (Table 2). For each plant extract studied, the inhibitory effect on P. grisea at 10% and 25% concentrations was significantly higher (P < 0.001) than 1% and 5% concentrations, except for B. pilosa and A. muricata where the difference between extract concentrations was not significant (P < 0.05). The extracts from C. arabica at 25% with 89.40% gave the highest radial growth inhibition of P. grisea, while the lowest (0.8%) radial growth inhibition was recorded on PDA amended with plant extracts from A. muricata at 5% (Table 2). It was generally observed that A. indica had fungicidal properties at all tested concentrations unlike other plant extracts tests. Our results are in close agreement to those reported by Amadioha [13] and Kamalakannan et al. [32] who observed that neem extract at different concentrations reduced the radial growth of P. grisea on PDA medium and incidences and severity of rice blast in-vivo. The effects of plant extracts increased with increasing concentration except extract from N. tabacum, which showed maximum mycelial inhibition at 10% concentration and thereafter, stimulated mycelial growth at a concentration of 25% (Table 2). Taiga and Friday [33] reported that N.

tabacum, A. vera and A. indica extracts contain alkaloids and flavonoids; while N. tabacum extract that was observed to be more fungi toxic at 10% concentration contained an additional component (tannins), which has been reported to be absent in both A. vera and A. indica extracts.

Seed extracts of processed C. arabica at 10% and 25% had the best inhibitory effect (81.12 and 89.40%), respectively, against P. grisea (Figure 1). They were followed by extracts from 10% N. tabacum (80.35%), 25% A. vera (79.45%), 25% C. coccineum flower (78.83%), 5% N. tabacum (78.33%), 25% C. arabica (77.52%), 10% A. vera (76.15%), 25% Z. officinalis (75.60%), 10% A. indica (74.90%), 10% C. coccineum (71.18 %), 5% A. indica (69.38%), 25% A. sativum (68.75%), 10% A. sativum (67.25%), 5% C. coccineum flower (61.73%) and 25% N. tabacum (61.15%) (Table 2). Similarly, Hajano et al. [34] evaluated garlic, neem and calatropis plant extracts and found that only garlic extract at higher dose successively reduced the infection caused by Magnaporthe oryzae on rice. It was also observed that neem extract at high dose 4 ml/15ml medium moderately inhibited the radial growth of M. oryzae.

Such results indicate that these plant extracts may have fungicidal properties that can be used as seed treatments for controlling P. grisea, the causal agent of rice blast. The antifungal properties of Z. officinalis on Aspergilus flavus, Aspergilus niger, Fusarium solani and Fusarium oxysporum on post-harvest yam (Dioscorea alata, Poir) has been reported by Yeni [35] . Other researchers also reported that Z. officinalis has been used for relief from arthritis, rheumatism, coughs, fever and infectious diseases [36] . The highest mycelial growth of P. grisea was favored by 5% B. pilosa (?10.92%) and 1% (?1.57%) (Table 2). The effect of N. tabacum on P. gri sea at 1%, 5% and 10% levels showed subsequent increase in percentage of fungal growth inhibition and decrease at 25% (Table 2). These results suggest that the optimal concentration of N. tabacum leaf extracts was around 10%.

Table 2. Percentage inhibition of radial growth of Pyricularia grisea on Potato Dextrose Agar mixed with plant extracts at various (v/v) concentrations.

Name of treatment	Plant extracts (% inhibition of mycelial growth)*			
	1%	5%	10%	25%
Control				
SDW	−3.23 h	−3.72 j	−3.47 j	−3.38 j
Apron Star® 42 WS	35.80 ab	50.5 d	82.1 a	89.8 a
Extracts source				
Zingiber officinalis	22.68 d	28.90 f	45.95 e	75.60 c
Azadirachta indica	37.23 ab	69.38 b	74.90 b	77.52 bc
Coffee arabica	34.08 b	50.58 d	81.12 a	89.40 a
Camellia sinensis	7.75 f	16.23 h	35.10 g	52.40 f
Nicotiana tabacum	26.82 c	78.33 a	80.35 a	61.15 e
Aloe vera	16.38 e	27.73 f	76.15 b	79.45 b
Datura stramonium	4.43 fg	23.15 g	39.25 f	49.10 g
Allium sativum	17.35 e	36.25 e	67.25 d	68.75 d
Annona muricata	0.80 gh	1.73 i	2.10 i	3.42 i
Chrysanthemum coccineum (leaf)	19.05 de	20.98 g	23.25 h	26.00 h
Bidens pilosa	−1.57 h	−10.92 k	−7.38 k	−5.43 j
Chrysanthemum coccineum (flower)	39.70 a	61.73 c	71.18 c	78.83 b
Mean	18.38	32.20	47.70	53.04
$LSD_{0.05}$	3.97	3.57	3.13	2.35
F test	***	***	***	***
Variance	7.72	6.28	4.83	2.72

*Values are means of four replicates in two separate experiments. Numbers followed by the same letters in a column are not significantly different at P = 0.05, using Duncan's Multiple Range Test; *** = highly significantly different (P < 0.001).

The shoot length of rice seedlings of cultivar Mwangaza were significantly different (P < 0.05) when treated with plant extracts, Apron Star® 42 WS and SDW. Seedlings of rice cultivar Mwangaza treated with Apron Star® 42 WS had the shortest shoot length (10.71 cm) compared with those treated with plant extracts and SDW (Table 3). There were no significant differences (P < 0.05) between the root length, dry weight, seedling growth rate and vigour index of cultivar Mwangaza for all treatments (Table 3). These results generally indicated that,

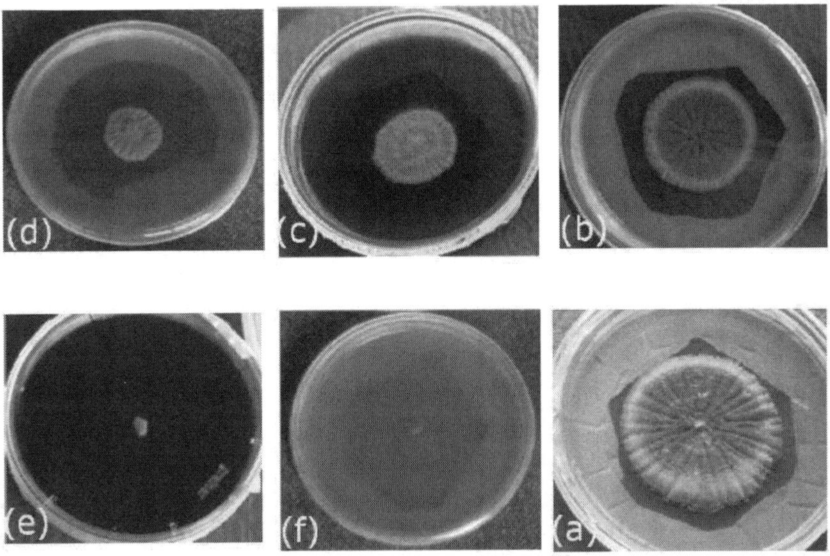

Figure 1. Growth of endophytic fungal mycelia on PDA amended with processed Coffee arabica: (a) = PDA amended with sterile distilled water (negative control), (b) = PDA amended with 1 %, (c) = PDA amended with 5 %, (d) = PDA amended with 10%, (e) = PDA amended with 25% and (f) = PDA amended with Apron star Apron Star® 42 WS (positive control) 21 days after incubation.

the plant extracts used were not phytotoxic to rice seedlings. These results are in agreement with those reported by Zida et al. [23] , who showed that Apron Star® 42 WS reduced seedling emergence, plant vigour and grain yield of pearl millet seeds. Seed treatments using ethanol extract and essential oil from Callistemon, Citrinus and Ocinum gratissimum improved the field parameters such as emergence and yield of tested rice varieties against Bipolaris oryzae [37] . The current findings suggest that leaf extracts from A. indica, A. vera, A. sativum, D. stramonium, C. sinensis, N. tabacum, Z. officinalis and processed C. arabica, can be used as seed treatments for the

Table 3. Effect of treatment with plant extracts on growth and development of rice plants in in-vitro using the between paper method.

Rice varieties	Treatment	Means of variables*					
		Germination (%)	Shoot length (cm)	Root length (cm)	Dry weight (g)	Seedling growth rate	Seedling vigour index
Supa	**Control**						
	SDW	96.00 a	10.50 ab	15.44 a	0.13 a	0.006 a	2369 ab
	Apron Star® 42 WS	95.00 a	9.55 b	14.57 a	0.10 b	0.004 b	2100 b
	Extracts source						
	Zingiber officinalis	97.00 a	10.44 ab	15.38 a	0.12 a	0.006 a	2345 ab
	Azadirachta indica	97.00 a	10.47 ab	16.05 a	0.13 a	0.006 a	2380 ab
	Coffee arabica	96.00 a	10.38 ab	15.47 a	0.12 a	0.005 ab	2362 ab
	Camellia sinensis	97.50 a	11.67 a	15.06 a	0.14 a	0.006 a	2435 a
	Nicotiana tabacum	97.00 a	10.09 b	16.18 a	0.13 a	0.005 ab	2424 ab
	Aloe vera	96.00 a	11.67 a	15.06 a	0.14 a	0.006 a	2375 ab
	Datura stramonium	96.50 a	10.62 ab	15.32 a	0.12 a	0.006 ab	2351 ab
	Allium sativum	96.00 a	10.65 ab	15.19 a	0.13 a	0.0060 ab	2376 ab
	Mean	96.40	10.60	15.37	0.12	0.006	2352
	$LSD_{0.05}$	2.21	1.15	1.83	0.01	0.001	280.0
	F test	ns	*	ns	***	ns	ns
	CV (%)	1.6	7.5	8.2	8.2	17.9	8.2
Saro-5	**Control**						
	SDW	95.50 a	11.46 b	14.02 a	0.13 ab	0.006a	2399 a
	Apron Star® 42 WS	91.50 b	9.76 c	13.48 a	0.10 b	0.005a	2011 a
	Extracts source						
	Zingiber officinalis	94.50 a	13.32 a	15.38 a	0.12 ab	0.006 a	2474 a
	Azadirachta indica	95.50 a	12.79 ab	15.53 a	0.11 ab	0.005 a	2448 a
	Coffee arabica	95.50 a	12.39 ab	14.00 a	0.14 a	0.006 a	2480 a
	Camellia sinensis	95.00 a	11.80 b	14.64 a	0.12 ab	0.006 a	2430 a
	Nicotiana tabacum	95.00 a	12.20 ab	14.00 a	0.12 ab	0.005 a	2383 a
	Aloe vera	94.50 a	11.97 b	14.76 a	0.13 a	0.006 a	2436 a
	Datura stramonium	96.00 a	11.99 b	15.36 a	0.13 a	0.006 a	2500 a
	Allium sativum	95.00 a	11.59 b	14.84 a	0.12 ab	0.006 a	2369 a
	Mean	94.80	11.93	14.60	0.12	0.006	2393
	$LSD_{0.05}$	2.56	1.19	2.15	0.02	0.001	264.6
	F test	ns	***	ns	ns	ns	ns
	CV%	1.9	6.9	10.2	12.0	10.9	7.7
Mwangaza	**Control**						
	SDW	95.50 a	11.87 ab	15.24a	0.11 a	0.005 a	2548 a
	Apron Star® 42 WS	94.50 a	10.71 b	12.82b	0.09 a	0.005 a	2200 b
	Extracts source						
	Zingiber officinalis	95.50 a	11.68 b	14.94 a	0.11 a	0.005 a	2472 a
	Azadirachta indica	95.00 a	12.75 a	15.02 a	0.10 a	0.005 a	2434 ab
	Coffee arabica	95.00 a	11.99 ab	14.33 a	0.11 a	0.005 a	2418 ab
	Camellia sinensis	95.00 a	11.74 ab	14.70 a	0.10 a	0.005 a	2499 a
	Nicotiana tabacum	95.00 a	11.63 b	14.53 a	0.12 a	0.005 a	2402 ab
	Aloe vera	95.50 a	11.74 ab	14.95 a	0.10 a	0.005 a	2462 a
	Datura stramonium	95.50 a	11.85 ab	15.48 a	0.11 a	0.005 a	2429 ab
	Allium sativum	95.50 a	11.83 ab	14.69 a	0.11 a	0.005 a	2441 ab
	Mean	95.20	11.78	14.67	0.10	0.005	2431
	$LSD_{0.05}$	2.72	0.92	1.38	0.02	0.001	221.9
	F test	ns	*	ns	ns	ns	ns
	CV%	2.0	5.4	6.5	14.3	12.3	6.3

*Values are means of four replicates. Numbers followed by the same letter in a column are not significantly different at P = 0.05, using Duncan's Multiple Range Test. *** = highly significantly different ($P < 0.001$); * = significantly different ($P < 0.05$); ns = not significantly different.

control of P. grisea in rice seeds and for improving rice seedling growth. These plant extracts are safe and eco-friendly on rice compared to synthetic chemicals such as Apron Star® 42 WS. More studies are therefore, needed to confirm current findings and to determine the most effective formulation against P. grisea.

Identification and characterization of the active compounds from currently tested plant extracts and their role in rice blast disease control is also needed. Research on the range of activity of plant extracts for control of other rice pathogens is recommended.

4. CONCLUSION

This study has revealed that plant extracts from processed coffee (C. arabica) at concentrations of 10% and 25% (w/v) had the highest inhibitory effect (81.12% and 89.40%, respectively) against P. grisea—the causal agent of rice blast disease. These results generally indicated that, the plant extracts used were not phytotoxic to rice seedlings. More studies are therefore needed to confirm the current findings and to determine the most effective formulation against P. grisea.

ACKNOWLEDGEMENTS

Judith Hubert, Robert B. Mabagala, Delphina P. Mamiro (AGRA) for financial support for this work.

REFERENCES

1. Ghazanfar, M.U., Habib, A. and Sahi, S.T. (2009) Screening of Rice Germplasm against Pyricularia oryzae, the Cause of Rice Blast Disease. Pakistan Journal Phytopathology, 21, 41-44.

2. Sukanya, S.L., Yamini, D. and Fathima, S.K. (2011) Eco-Friendly Management of Pyricularia oryzae, the Causal Agent of Blast of Paddy. Journal of Current Botany, 2, 46-49.

3. TeBeest, D.O., Guerber, C. and Ditmore, M. (2007) Rice Blast. Journal of Plant Disease, 10,109-113.

4. International Rice Research Institute (IRRI) (2010) Magnaporthegrisea: Rice Blast.http://www.metapathogen.com.visited

5. Luzi-Kihupi, A., Zakayo, J.A., Tusekelege, H., Mkuya, M., Kibanda, N.J.M., Khatib, K.J. and Maerere, A. (2009) Mutation Breeding for Rice Improvement in Tanzania. Food and Agriculture Organization, United Nations, Rome.

6. Kato, K. (2007) Development of a Major Rice Cultivation Area in the Kilombero Valley, Tanzania. African Study Monographs, 36, 3-18.

7. Mossini, S.A.G., Oliveira, K.P. and Kemmelmeier, C. (2004) Inhibition of Patulin Production by Penicilliumexpansum Cultured with Neem (Azadirachtaindica) Leaf Extracts. Journal of Basic Microbiology, 44, 106-113.

8. Mwilene, F.E., Oikeh, S.O., Agunibiade, T.A., Oladimeji, O., Ajayi, O., Sie, M., Gregorio, G.B., Togola, A. and Toure, A.D. (2008) Growing Lowland Rice. In: Production Handbook, Africa Rice Center (WARDA), Bouaké.

9. International Rice Research Notes (IRRN) (1993) Treating Rice Seeds with Fungicides and Antagonists to Control Seed-Borne Diseases.

10. Mbodi, Y., Gaye, S. and Diaw, S. (1986) The Role of Tricyclazole in Rice Protection against Blast and Cultivar Improvement. Journal of Parasitica, 43, 187-198.

11. Khan, Z.S. and Nasreen, S. (2010) Phytochemical Analysis, Antifungal Activity and Mode of Action of Methanol Extracts from Plants against Pathogens. Journal of Agricultural Technology, 6, 793-805.

12. Peter, A.C.O. (2005) Non-Pesticide Methods for Controlling Diseases and Insect Pests. Integrated Pest Management Expert and Food and Agriculture Organization, Bangkok. [Paper reference:1]

13. Amadioha, A.C. (2000) Controlling Rice Blast in Vitro and in Vivo with Extracts of Azadirachta indica. Journal of Crop Protection, 19, 287-290.

14. Jabeen, R. (2006) Evaluation of Botanicals against Xanthomonas oryzae pv. oryzae Using Bacterial Leaf Blight Disease of Rice and Characterization of Bioactive Compounds. MSc. Dissertation, Quaid-i-Azam University, Islamabad.

15. Lalitha, V., Raveesha, K.A. and Kiran, B. (2010) Antimicrobial Activity of Solanum torvum Swart. against Important Seed Borne Pathogens of Paddy. Iranica Journal of Energy & Environment, 1, 160-164.

16. Enikuomehin, O.A. (2005) Cercospora Leaf Spot Disease Management in Sesame (Sesamum indicum L.) with Plant Extracts. Journal of Tropical Agriculture, 43, 19-23. [Paper reference:1]

17. Gurjar, M.S., Ali, S., Akhtar, M. and Singh, K.S. (2012) Efficacy of Plant Extracts in Plant Disease Management. Journal of Agricultural Sciences, 3, 425-433.

18. Mathur, S.B. and Kongsdal, O. (2004) Common Laboratory Seed Healthy Testing Methods for Detecting Fungi. 2nd Edition, International Seed Testing Association (ISTA), Bassersdorf.

19. Harlapur, S.I., Kulkarni, M.S., Wali, M.C. and Kulkarni, S. (2007) Evaluation of Plant Extracts, Bio-Agents and Fungicides against Exserohilum turcicum (Pass.) Leonard and Suggs. Causing Turcicum Leaf Blight of Maize. Journal of Agricultural Sciences, 20, 541-544.

20. Akinbode, O.A. and Ikotun, T. (2008) Evaluation of Some Bio-Agents and Botanicals in in Vitro Control of Colletotrichuwm destructivum. African Journal of Biotechnology, 7, 868-872.

21. Hasan, M.M., Chowdhury, S.P., Alam, S., Hossain B. and Alam, M.S. (2005) Antifungal Effects of Plant Extracts on Seed-Borne Fungi of Wheat Seed Regarding Seed Germination, Seedling Healthy and Vigor Index. Journal of Biological Sciences, 8, 1284-1289.

22. Nduagu, C., Ekefan, E.J. and Nwankiti, A.O. (2008) Effect of Some Crude Plant Extracts on Growth of Colletotrichum capsici (Synd) Butler and Bisby, Causal Agent of Pepper Anthracnose. Journal of Applied Biosciences, 6, 184-190.

23. Zida, E.P., Sereme, P., Leth, V. and Sankara, P. (2008) Effect of Acacia gourmaensis A. Chev and Eclipta alba (L.) Hassk. on Seed Health, Seedling Vigour and Grain Yield of Sorghum and Pearl Millet. Asian Journal of Plant Pathology, 2, 40-47.

24. Mamiro, D.P. and Royse, D.J. (2004) Laboratory Efficacy of Selected Fungicides and Rhododendron catawbiense Leaf Extracts on the Growth of Vericillium fungicola. Acta Edulis Fungi, 12, 390-396.

25. Kamalakannan, A. and Shanmugam, V. (2005) Controlling Downy Mildew of Maize Caused by Peronosclerospora sorghi by Prosopis chilensis, Azadirachta indica and Foliar Sprays of a Talc-Based Formulation of a Pseudomonas fluorescens. Plant Pathology, 18, 115-118. [Paper reference:1]

26. Ogbebor, N. and Adekunle, A.T. (2005) Inhibition of Conidial Germination and Mycelial Growth of Corynespora cassiicola (Berk and Curt) of Rubber (Hevea Brasiliensis Muell. Arg.) Using Extracts of Some Plants. African Journal of Biotechnology, 4, 996-1000. [Paper reference:1]

27. Namai, T. and Ehara, Y. (1986) Studies on Variation in Virulence of Rice Blast Fungus, Pyricularia oryzae Cavara. Journal of Agricultural Research, 36, 3-4.

28. Chala, A., Brurberg, M.B. and Tronsmo, A.M. (2010) Incidence and Severity of Sorghum Anthracnose in Ethiopia. Plant Pathology, 9, 23-30.

29. International Seed Testing Association (ISTA) (2008) International Rules for Seed Testing. Bassersdorf. [Paper reference:4]

30. Mbega, E.R., Mortensen, C.N., Mabagala, R.B. and Wulff, E.G. (2012) The Effect of Plant Extracts as Seed Treatments to Control Bacterial Leaf Spot of Tomato in Tanzania. Journal of General Plant Pathology, 78, 277-286.

31. Chhetri, S. (2009) Identification of Accelerated Aging Conditions for Seed Vigor Test in Rice (Oryza sativa L.). Ph.D. Thesis, Thawatchai Teekachunhatean, Nakhon Ratchasima. [Paper reference:1]

32. Kamalakannan, A., Shanmugam, V., Surendran, M. and Srinivasan, R. (2001) Antifungal Properties of Plant Extracts against Pyricularia grisea, the Rice Blast Pathogen. Indian Journal of Phytopathology, 54, 490-492.

33. Taiga, A. and Friday, E. (2009) Variations in Phytochemical Properties of Selected Fungicidal Aqueous Extracts of Some Plant Leaves in Kogi State, Nigeria. American-Eurasian Journal of Sustainable Agriculture, 3, 407-409.

34. Hajano, J.U., Lodhi, A.M., Pathan, M.A., Khanzada, M.A. and Shah, G.S. (2012) In Vitro Evaluation of Fungicides, Plant Extracts and Bio-Control Agents against Rice Blast Pathogen Magnaporthe oryzae Couch. Pakistan Journal of Botanical, 44, 1775-1778. [Paper reference:1]

35. Yeni, I.J. (2011) Evaluation of Antifungal Effects of Extracts of Allium sativum and Nicotiana tabacum against Soft Rot of Yam (Dioscorea alata). Journal of Agricultural Research, 3, 1-5.

36. Obongoya, B.O., Wagai, S.O. and Odhiambo, G. (2009) Fungitoxic Properties of Four Crude Plant Extracts on Fusarium oxysporum Schl. F. sp. Phaseoli. African Journal of Food Agriculture Nutrition and Development, 9, 1652-1666.

37. Nguefack, J., Torp, J., Leth, V., Dongmo, J.B., Fotio, D. and Amvam Zollo, P.H. (2007) Effects of Plant Extracts and Chemical Fungicide in Controlling a Rice Seed-Borne Fungus under Laboratory and in Irrigated Cropping System in Ndop-Cameroon. African Journal of Crop Science, 8, 791-796.

Chapter 6

Usage of the Relationship Between the Application Rates of the Active Ingredient of Fungicides and Their Residue Levels in Mature Apples to Creating a Coherent System of Mrls

Stanisław Sadło[1] • Stanisław Walorczyk[2] • Przemysław Grodzicki[3] • Bartosz Piechowicz[1]

[1.] *Institute of Applied Biotechnology and Basic SciencesUniversity of RzeszówKolbuszowaPoland*

[2.] *Department of Pesticide Residue ResearchInstitute of Plant Protection – National Research InstitutePoznańPoland*

[3.] *Department of Animal Physiology, Faculty of Biology and Environment ProtectionNicolaus Copernicus UniversityToruńPoland*

ABSTRACT

Setting of maximum residue levels (MRLs; tolerances in the USA) in crops requires a big amount of data concerning residues from a number of supervised field trials for each pesticide/crop combination. This task is time-consuming, costly and fairly complicated. Therefore, we initiated a study on the utilization of the interpolation method in supporting and facilitating of assessing the value of the MRLs which are now being in use, and in creating their consistent system in the future. A mathematical formula for predicting the initial pesticide residue level in mature apples was developed by establishing the relationship between application rates and residue levels of the pesticide active ingredients in mature apples. This dependence was described by a linear equation $R_0 = 1.2593 \times D$, with coefficient of determination $r^2 = 0.984$, where D is an application rate of a given substance. This relationship makes it possible to predict a residue level of other substances of fungicidal activity. Thus, residue levels (both predicted by the formula and obtained in the frame of Polish National Monitoring Program) of fungicides now used in apple orchards, in mature apples were typically lower than those of the statutory MRLs. Bearing in mind that the European Union policy is directed towards substantial reduction in pesticide use, the MRLs might be reconsidered in order to be set at lower values.

Keywords: Application rate Mature apples Maximum residue levels Pesticide residues

INTRODUCTION

Among the food safety hazards for the consumers, pesticide residues, some bacterial pathogens (e.g. *Salmonella enterica*) and foodborne viruses (e.g. norovirus) have been identified as three most important risk factors [4]. In order to safeguard consumers' health and to promote principles of good agricultural practice (GAP) in the use of pesticides, maximum residue levels (MRLs; tolerances in the USA) of pesticides have been set by all EU Member States applying the same evaluation procedures and authorization criteria in order to put a plant protection product (PPP) on the market. These MRLs represent the highest concentrations of pesticides (expressed in mg of active ingredient per kg) which are legally permitted in food commodities, and they are interpreted as the highest residue of a given substance which may be found if a pesticide, registered and authorized, is applied according to label and, therefore, also to principles of GAP [8]. So defined MRLs can be established only when the residues in a crop resulting from particular use pattern of the pesticide meet the public health risk assessment criteria [8, 14,19]. Despite international discrepancies between food safety regulations, compliance with MRLs is still an essential prerequisite in trade of food and agricultural products [6, 23, 24].

The residue levels of pesticides occurring in crops at harvest time are dependent on many factors and are susceptible to influence of parameters such as a spray equipment used, a spray quality and conditions, crop management, crop varieties, practices of growing, growth stages at time of application and weather conditions [8]. But the most important determining factors are application rate of the active ingredient (AI) per hectare and an interval between the last application and the harvest (pre-harvest interval PHI). It appears fairly obvious that relatively high residues may be expected when pesticides are applied at high doses later in growing season when plant growth rate is slow [17].

For setting MRLs, residue data obtained from a number of supervised field trials are required for each pesticide/crop combination. This task is time-consuming, costly and fairly complicated to perform [2, 3, 8, 16]. However, development of a consistent approach for supporting and facilitating estimation of the MRLs based on effective application rate (dose) of the AI appears to be feasible. For this reason, we initiated the field trials to investigate whether the interpolation method may be efficient in predicting of the residue levels of fungicides now used in apple orchards against diseases that potentially may develop during fruit storage. Apples are particularly interesting study object, since these fruits are heavily exposed to pesticide contamination due to numerous pesticide treatments to whom they are subjected, and because they are the most highly consumed fruits (along with oranges) in the European Union countries and the USA.

MATERIALS AND METHODS

Field trials

The field trials were carried out in a commercial orchard which was located near the Kraśnik town (Lublin Province, south-eastern Poland) in 2011–2012. The orchard is specialized in providing apples for baby food production and was certified in compliance with Integrated Pest Management (IPM) system by the Main Inspectorate of Plant Health and Seed Inspection (PIORiN—Państwowa Inspekcja Ochrony Roślin i Nasiennictwa).

Trial 1: To protect apples against fungal diseases that develop during the cold storage, apple trees of Gloster variety were sprayed with Switch 62.5 WG, a commercial product containing 375 g kg^{-1} of cyprodinil and 250 g kg^{-1} of fludioxonil in the form of water dispersible granules (WG), at a dose of 0.8 kg ha^{-1}. Switch 62.5 WG, which is a mixture of compounds belonging to anilinopyrimidine and phenylpyrrole chemical groups, exhibits a protective and systemic activity against infections of fruits and vegetables by Botrytis cinerea (grey mould).

Trial 2: One week before harvesting mature apples, and submitting them to the storage room, apple trees of Lobo variety were sprayed with Zato 50 WG, at a dose of 0.2 kg ha^{-1}. Trifloxystrobin, the AI of this plant protection product, belongs to the family of strobilurin fungicides and exhibits systemic mode of action.

Trials 3 and 4: These two field trials were carried out on the apple trees of Gloster and Lobo varieties, which were sprayed with fungicides Merpan 80 WG and Captan 80 WG (in both cases, the active ingredient was captan) at a dose of 1.9 kg ha^{-1}. According to the label, Merpan 80 WG or 50 WP is recommended to control apple scab (*Venturia inaequalis*). Preparations containing captan (and folpet belonging to the same phthalimide group) are the fungicides commonly used to control the mentioned disease during the whole vegetative period.

Trial 5: Apple trees of Golden Delicious variety were sprayed with the Bellis 38 WG (25.5 % boscalid and 12 % pyraclostrobin) using 0.8 kg of the fungicide per ha. Bellis 38 WG is composed of two AIs: pyraclostrobin belonging to strobilurin group, and boscalid, preventive fungicide with curative action belonging to the anilide chemical group.

Trial 6: Similarly to the field trial 2, apple trees of Gloster variety were sprayed with Zato 50 WG, but at a lower dose of 0.15 kg ha^{-1}.

Sampling

Each time, sampling was made at the next day after pesticide application (approximately after 12 h). Each laboratory sample consisted of eight apples collected from randomly selected trees. Of each field trial, four laboratory samples were taken, each of them from one randomly selected row of the apple trees. Fruit samples were transported to the laboratory where they were homogenized, and then analytical portions of known weight (each of them weighing about a hundred grams) were used for the pesticide residue analysis.

Pesticide residue analysis

Reagents and standards: acetone and dichloromethane of HPLC grade were obtained from Sigma-Aldrich Sp. z o.o., Poznań, Poland. Petroleum ether (Chempur, Piekary Śląskie, Poland) and diethyl ether (POCH, Gliwice, Poland) were distilled from glass before use. Sodium sulphate(VI) (POCH, Gliwice, Poland) was heated at 550 °C for 7 h. Florisil (Sigma-Aldrich Sp. z o.o., Poznań, Poland) was activated by heating at 130–135 °C for 7 h and stored in a desiccator before use. Certified pesticide analytical standards were obtained from

Ehrenstorfer (Augsburg, Germany). Stock solutions of approximately 1000 μg mL^{-1} were prepared in acetone and stored at −17 °C of which intermediate concentration standards (10 μg mL^{-1}) were prepared by dilution with acetone (stored at 4 °C). Working standard mixtures were obtained by diluting the intermediate concentration solutions with appropriate volumes of petroleum ether and stored at 4 °C.

Sample preparation: the sample preparation was carried out according to the analytical method which was previously described in detail [18, 21]. Briefly, an analytical portion (about 100 g) was homogenized for 2 min with 150 mL of acetone. The extract was filtered, and then an aliquot of the filtrate (equivalent to 20 g of the analytical portion) was taken for further processing. After addition of 100 mL of 2.5 % aqueous solution of sodium sulphate(VI), it was partitioned sequentially liquid–liquid with 20, 10 and 10 mL dichloromethane. The combined extracts were evaporated to dryness and dissolved with 10 mL petroleum ether. An aliquot of 5 mL was transferred into a glass column which was previously packed with 1.1 g activated florisil and 4.5 g anhydrous sodium sulphate(VI). The residues were eluted with 70 mL diethyl ether/petroleum ether (3:7, *v*/*v*) and 70 mL acetone/petroleum ether (3:7, *v*/*v*). The combined extracts were evaporated and diluted with petroleum ether. The concentration of matrix in the final extract was 1 g per mL.

Gas chromatographic determination: a gas chromatograph model 7890 (Agilent Technologies, Palo Alto, CA, USA) was used. The instrument was equipped with a micro-electron capture (μEC) and nitrogen-phosphorus (NP) detectors. The HP-5 MS Ultra Inert capillary column 30 m × 0.32 mm × 0.25 μm film thickness (Agilent Technologies, Palo Alto, CA, USA) was used. Sample extracts of 2 μl were injected splitless at 250 °C. The column temperature was ramped from 100 to 180 °C at 10 °C min^{-1} (held for 4 min), then ramped to 220 °C at 3 °C min^{-1} (held for 15 min), then ramped to 260 °C at 10 °C min^{-1} (held for 11 min). Cyprodinil and fludioxonil were determined by using the nitrogen-phosphorus detector (NPD) which was operated at

300 °C while the gas flows were as follows: hydrogen 3 mL min^{-1}, air 60 mL min^{-1} and makeup (nitrogen) 10 mL min^{-1}. Boscalid, captan, piraclostrobin and trifloxystrobin were determined by using the micro-electron capture detector (μECD) which was operated at 280 °C, and the makeup gas flow (nitrogen) was 30 mL min^{-1}. Agilent Technologies ChemStation Rev.B04.03 software was used for instrument control, data acquisition and evaluation.

Pesticide residue estimation

Residues of the tested substances were determined by gas chromatographic analysis, and their values were expressed in mg of AI per kg of apples. Average residue levels ($n = 4$), standard deviations and coefficients of variations (CV)/relative standard deviations (RSDs) of the obtained values were calculated. Subsequently, the relationship between the two variables: average residue level (y) and application rate (x) of the tested substances was established in the form of linear equation of $y = ax$, based on the assumption that this relationship is linear, and the straight line goes through the point (0,0) on graph (y-intercept = 0).

RESULTS AND DISCUSSION

Method validation study

To ensure quality of the analytical results, the method was validated before the use to determine the fungicide residues in the apple samples (Table 1). The recovery of the fungicides was determined for samples of untreated apples which were spiked with the target compounds on two concentration levels ($n = 5$). The average recoveries were satisfactory because they were compliant with the European Union criteria (SANCO/12495/2011) which stipulate the average recoveries in the range 70–120 % with corresponding RSD less or equal 20 % (http://ec.europa.eu/food/plant/plant_protection_products/guidance_documents/docs/qualcontrol_en.pdf).

Table 1. Method performance characteristics: limit of quantification (LOQ), linearity parameters, recovery, relative standard deviation (RSD) and expanded uncertainty values (U) of fungicide analysis, in apples

Pesticides	LOQ (mg kg^{-1})	Linearity range (mg kg^{-1})	R^2	Spiking level (mg kg^{-1})	Recovery ± RSD (%)	U (%) (k = 2)
Boscalid	0.01	0.008–0.211	0.999	0.011	86.5 ± 2.3	10
				0.211	91.7 ± 0.9	
Captan	0.01	0.008–0.210	0.997	0.011	94.1 ± 15.1	35
				0.210	119.1 ± 7.3	
Cyprodinil	0.01	0.011–0.269	0.999	0.013	95.5 ± 3.4	10
				0.269	94.8 ± 2.8	
Fludioxonil	0.02	0.018–0.454	0.995	0.023	80.0 ± 14.1	35
				0.454	71.1 ± 3.3	
Piraclostrobin	0.01	0.008–0.201	0.998	0.010	118.2 ± 13.7	35
				0.201	119.7 ± 7.7	
Trifloxystrobin	0.01	0.007–0.171	0.999	0.009	98.1 ± 2.8	8
				0.171	93.0 ± 1.8	

The limit of quantification (LOQ) was defined as the lowest spiking level (rounded to one significant number) for which the validation criteria were satisfied and it was equal 0.01 mg kg^{-1} for boscalid, captan, cyprodinil, piraclostrobin and trifloxystrobin, and 0.02 mg kg^{-1} for fludioxonil. Excellent linearity with the coefficients of determination (R^2 > 0.99) was achieved for the studied pesticides when using matrix-matched standards.

Measurement uncertainty of the method was estimated based on the "top-down" approach by using the recovery and precision data obtained in the validation study [22]. The expanded uncertainty was calculated individually for each pesticide as twice the value of the uncertainty (k = 2, confidence level 95 %), and it ranged from 8 to 35 % (22 % on average), which was distinctively less than a maximum default value of ±50 % recommended by the European Union guidelines SANCO/12495/2011 (http://ec.europa.eu/food/plant/plant_protection_products/guidance_documents/docs/qualcontrol_en.pdf).

Field trials

The post-harvest treatment of the apple fruits (e.g. dipping) against the storage diseases of fungal origin is not allowed in Poland. In order to keep fruit quality and to provide effective protection against storage diseases, 7–14 days before harvest (pre-harvest interval PHI), growers spray apples that are still on the trees with fungicides, to control pathogens responsible for developing these diseases. Because these treatments are performed in the orchard as the last ones before harvest, they generate the highest possible residue levels of the applied active ingredients (AIs).

To determine the overall relationship between application rate of the AI of plant protection product and its residue level in mature apples, six field trials were carried out on four varieties of dessert apples (Elise, Gloster, Golden Delicious and Lobo) using five plant protection products (Bellis 38 WG, Captan 80 WG, Merpan 80 WG, Switch 62.5 WG and Zato 50 WG) containing a total six different AIs (Table 2). The studied AIs were: captan, pyraclostrobin, trifloxystrobin, boscalid, cyprodinil and fludioksonil, belonging to four different chemical groups.

Table 2. Data on supervised field trials and the average residue levels of active ingredients, just after application of the plant protection product

Trial nos.	Apple variety	Fungicide formulation (WG)	Application rate of product (kg or L ha^{-1})	Active ingredient (AI)	Application rate of AI (kg ha^{-1})	Average residue level (mg kg^{-1})
1	Gloster	Switch 62.5	0.800	Cyprodinil	0.300	0.236 ± 0.064
				Fludioxonil	0.200	0.294 ± 0.039
2	Lobo	Zato 50	0.200	Trifloxystrobin	0.100	0.126 ± 0.016
3	Gloster	Merpan 80	1.900	Captan	1.520	2.090 ± 0.545
4	Elise	Captan 80	1.900	Captan	1.520	1.765 ± 0.440
5	Golden Delicious	Bellis 38	0.800	Boscalid	0.202	0.236 ± 0.128
				Pyraclostrobin	0.102	0.092 ± 0.051
6	Gloster	Zato 50	0.150	Trifloxystrobin	0.075	0.106 ± 0.031

Captan (trials 3 and 4) is a non-systemic phthalimide fungicide which is used to control many fungal diseases of fruit, mainly in apple production. Captan (Captan 80 WG, Merpan 80 WG) has a protective activity, and the main product of its degradation is tetrahydrofthalimid (THPI), which was also detected in some apples intended for manufacturing baby foods (unpublished data) and during apple processing [1]. Folpet (Folpan 80 WG) is a chemical analogue of captan, and it is also registered for use in apple orchards in Poland. Similarly to captan, it undergoes decomposition to a phthalimide degradation product [5].

In trials 2, 5 and 6, pyraclostrobin and trifloxystrobin were used. Both compounds belong to the strobilurin group. They have systemic and protective activity [15]. In Poland, formulations containing these substances are often used. Compared with captan or folpet, these substances protect the fruit at lower doses not exceeding 100 g ha^{-1}. Another studied fungicides were: boscalid (an anilide fungicide of systemic action), cyprodinil (an anilinopyrimidine fungicide of systemic action) and fludioxonil (a phenylpyrrole fungicide) [7]. These three fungicides are effective at doses that can be classified as medium (200 g ha^{-1}).

All the fungicides used in the field trials were applied in accordance with the label recommendations, and their application rates covered a wide range of doses (*D*), depending on the active ingredient. The lowest dose was used in the case of strobilurin fungicides (0.075–0.102 kg ha^{-1}), somewhat higher in the case of boscalid, cyprodinil and fludioxonil (0.200–0.300 kg ha^{-1}), and the highest for captan (1.52 kg ha^{-1}).

Application rates versus residue levels of fungicides

Data on field trials and residue levels of boscalid, captan, cyprodinil, fludioxonil, pyraclostrobin and trifloxystrobin which have been found in mature apples in the following days after their applications are summarized in Table 2. In estimation of the obtained results, we assumed that the relationship between the dose of any AI of plant protection product and its average residue level in mature apples may be well approximated by a linear function, and thus, based on obtained results, such correlation had been expressed by the following equation (see also Fig. 1):

R0=1.2593×D;Coefficient of determination r2=0.9846,R0=1.2593×D;Coefficient of determination r2=0.9846,

where R_0 is the average residue level (mg kg^{-1}) of a given substance which may be found in mature apples just after its application at a *D* dose (kg ha^{-1}), and the numeric value of 1.2593 represents residue level of any substance if it is applied at a dose of 1 kg ha^{-1}.

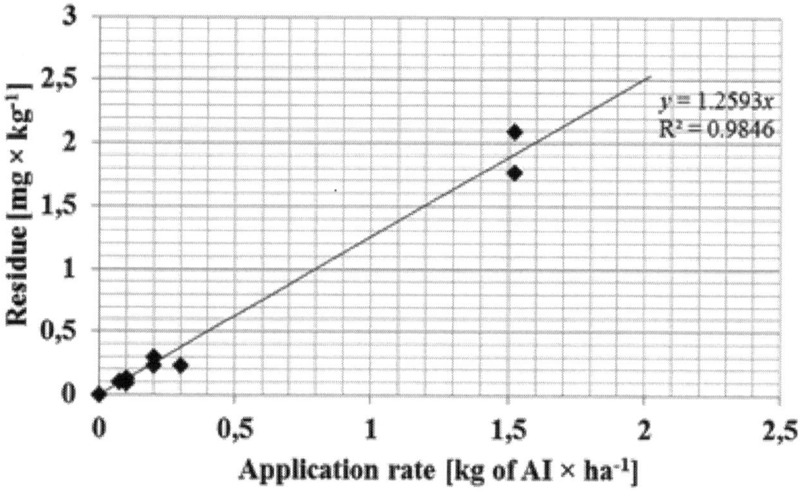

Figure 1. Initial residue level (*y*-axis) versus the application rate (*x*-axis) of the studied fungicides

As can be seen in Table 3, this mathematical formula makes possible the calculation of the average residue level of any fungicide now in use in the apple orchards, as well for each substance possessing fungicidal properties for the future introduction to the plant protection programs. Moreover, knowing the initial residue level (R_0) of a given substance and constant rate (k) of its exponential dissipation, determined independently, we may predict its concentration at any t-time, according to the general formula: $R_t = R_0 \times e^{-kt}$ [9] and then

Rt=1.2593×D× e−kt.Rt=1.2593×D× e−kt.

The above formula, moreover, makes it possible to estimate residue level (R_t = PHI) of a given substance just after a fixed pre-harvest interval (PHI) and to compare it with the monitoring data, to ensure the correctness of the **pesticide application in crop protection.**

Table 3. Predicted initial residue levels ($R_0 = 1.2593 \times D$) of AIs of fungicides against apple scab (*Venturia inaequalis*) in ripe apples, just after their application according to the label directions (D—application rate and R_0—predicted initial residue level) in comparison with the MRLs now in force

Chemical group fungicide	PHI (days)	AI	D^a (kg ha^{-1})	R_0^a (mg kg^{-1})	MRL (mg kg^{-1})
Anilinopyrimidines					
Chorus 50 WG	ND	Cyprodinil	0.15	0.19	1
Mythos 300 SC	14	Pyrimethanil	0.3 (0.45)	0.38 (0.57)	5
Vision 250 SC	30	Pyrimethanil	0.2 (0.3)	0.25 (0.38)	5
Quinones					
Delan 700 WG	21	Dithianon	0.35 (0.525)	0.44 (0.66)	3
Tercel 16 WG	35	Dithianon	0.24 (0.3)	0.30 (0.38)	3
Ventop 350 SC	14	Dithianon	0.35 (0.525)	0.44 (0.66)	3
Benzimidazoles					
Topsin M 500 SC	14	Thiophanate-methyl	0.75	0.94	0.5
Phthalimides					
Captan 50 WP	7	Captan	1.5	1.89	3
Captan/Merpan 80 WG	7	Captan	1.52 (2.24)	1.91 (2.82)	3
Kaptan Plus 71.5 WP	7	Captan	1.4	1.76	3
Kaptan zawiesinowy 50 WP	7	Captan	1.5	1.89	3
Folpan 80 WG	20	Folpet	2.4	3.02	3
Shavit F 72	20	Folpet	1.05 (1.4)	1.32 (1.76)	3
Guanidine derivatives					
Carpene 65 WP	14	Dodine	0.975	1.23	5
Syllit 65 WP	14	Dodine	0.65 (1.46)	0.82 (1.84)	5
Strobilurins					
Ardent 500 SC	28	Kresoxim-methyl	0.1	0.13	0.2
Discus 500 WG	28	Kresoxim-methyl	0.1	0.13	0.2
Signum 33 WG	14	Pyraclostrobin	0.054	0.06	0.3
Tercel 16 WG	35	Pyraclostrobin	0.08 (0.1)	0.10 (0.13)	0.3
Zato 50 WG	14	Trifloxystrobin	0.075	0.09	0.5
Bellis 38 WG	14	Pyraclostrobin	0.102	0.13	0.3
Anilides					
Signum 33 WG	14	Boscalid	0.214	0.27	2
Bellis 38 WG	14	Boscalid	0.202	0.25	2
Triazoles					
Bumper 250 SC	14	Propiconazole	0.075	0.09	0.15
Capitan 400 EC	14	Flusilazole	0.03	0.04	0.02
Difo 250 SC	ND	Difenoconazole	0.05	0.06	0.5
Domark 100 EC	ND	Tetraconazole	0.06	0.08	0.3
Indar 5 EW	28	Fenbuconazole	0.035	0.04	0.4
Kaptan Plus 71.5 WP	7	Triadimenol	0.03	0.04	0.2
Riza 250 EW	14	Tebuconazole	0.125	0.16	1
Score 250 EC	14	Difenoconazole	0.05	0.06	0.5
Shavit 72 WG	20	Triadimenol	0.03 (0.04)	0.04 (0.05)	0.2
Sparta 250 EW	14	Tebuconazole	0.125	0.16	1
Suplo 250 EC	14	Difenoconazole	0.05	0.06	0.5
Systemic 125 SL	7	Myclobutanil	0.056 (0.075)	0.07 (0.09)	0.5
Talent 240 EC	14	Myclobutanil	0.06 (0.072)	0.08 (0.09)	0.5
Troja 250 EW	14	Tebuconazole	0.125	0.16	1
Vision 250 SC	30	Fluquinconazole	0.05 (0.075)	0.06 (0.09)	0.1

[a] Application rates (D), and thus predicted initial residue level (R_0) added in brackets relate to periods of high infection pressures

Predicted residue levels of fungicides just after treatment versus their MRLs now in force

The reduction of yield losses caused by pests and pathogenic fungi is the major challenge for agricultural production [13]. Ideally, pesticides should maintain the pest population at lower level than that causing economically unacceptable damage or loss, but without leaving residues in the edible parts of crops [10]. This is possible very rarely, and for this reason, legally, maximum residue levels (MRLs; "tolerances" in the USA) must be established for particular combination of pesticide and commodity. These MRLs are the upper permissible levels of pesticide residues and are set in order to ensure the lowest possible (!) consumer exposure according to the European Union legislation framework Regulation (EC) No 396/2005 (http://eur-lex.europa.eu/LexUriServ/LexUriServ.do?uri=OJ:L:2005:070:0001:0016:en:PDF).

Table 3 summarizes plant protection products recommended for control of diseases of fungal origin in apple orchards in Poland and their application rates along with MRLs currently in force in the European Union. As seen, the expected initial pesticide residue levels were all well below European MRLs, and therefore, they do not indicate any possibility of violation of regulatory limits. In general, our study indicates a significant discrepancy between MRLs and residue levels (R_0) immediately after treatments as calculated on the basis of the established relationship. We found that good consistency between the two values occurs in the case of kresoxim-methyl (MRL = 0.2 vs. R_0 = 0.13 mg kg^{-1}), captan and folpet (MRL = 3 vs.R_0 < 3 mg kg^{-1}) as well as for some triazoles, namely for fluquinconazole (MRL = 0.1 vs.R_0 = 0.09 mg kg^{-1} in periods of the high infection pressure) and propiconazole (MRL = 0.15 vs.,R_0 = 0.09 mg kg^{-1}). However, even for those few substances, after passing PHI period their residues will be also significantly below the MRLs.

The MRLs of other substances belonging to the triazole group (tebuconazole, myclobutanil, difenoconazole, triadimenol, fenbuconazole, tetraconazole), strobilurines (pyraclostrobin and trifloxystrobin), anilinopyrimidines (cyprodinil and pirimethanil) as well as the dodine and dithianon seem to be overestimated from 2 to 13 times. Furthermore, the discrepancies between MRLs and the final residue levels would be even greater, when taking into account their actual declines during the PHIs.

The proposed model of estimation of the residue level of fungicides in apples may facilitate not only setting MRLs for those fungicides and creation of their coherent system but also assessing the level of pesticide residues found in fruit and vegetables in the frame of National Monitoring Programs. Our results of the residue calculations in apples just after the treatment correspond well with the results of numerous data of pesticide residue monitoring. For example, in 2008, 141 of apple samples have been tested in Poland. Among the fungicides, the most frequently detected was captan (48.2 % samples), and the highest residue level was 1.9 mg kg^{-1} (the MRL value is 3 mg kg^{-1}), while trifloxystrobin residue was found 0.02 mg kg^{-1} (the MRL value is 0.5 mg kg^{-1}). The highest residue levels of cyprodinil, difenoconazole, dithiocarbamates, flusilazole, folpet and pyrimethanil were: 0.06, 0.08, 0.66, 0.09, 0.33 and 0.35 mg kg^{-1}, respectively, and all of them were well below the applicable MRLs [12]. These findings are also consistent with the results of other monitoring surveys conducted in Poland [11] and Czech Republic [20].

Pesticide MRLs provide an enforcement tool to ensure compliance with the registered pesticide labels and reassure that food products that move in commerce are safe for consumers. According to current registration practices, as described in Regulation (EC) No 1107/2009, pesticide MRLs are set based on data derived from supervised field trials (http://eur-lex.europa.eu/legal-content/EN/TXT/PDF/?uri=CELEX:32009R1107&from=EN). On the other hand, according to Regulation (EC) No 396/2005 "MRLs should be set at the lowest achievable level consistent with good agricultural practice for each pesticide with a view to protecting vulnerable groups such as children and the unborn" (http://eur-lex.europa.eu/LexUriServ/LexUriServ.do?uri=OJ:L:2005:070:0001:0016:en:PDF). Also, bearing in mind that the European Union pesticide use policy, as reflected in Directive 2009/128/EC, is directed towards significant reductions in pesticide use (http://eur-lex.europa.eu/LexUriServ/LexUriServ.do?uri=OJ:L:2009:309:0071:0086:en:PDF), we proposed that certain MRLs could be reconsidered and set at lower values based on pesticide residue monitoring data from surveys on crops produced under typical commercial conditions. It must be highlighted that applying the proposed approach neither mean a significant reduction in the number of initial field trials nor eliminate the need for zonal evaluation in the pesticide registration process but could be very useful for reviewing the existing MRLs, setting temporary MRLs and setting MRLs for pesticide residues in minor crops.

CONCLUSIONS

We proposed a simplified mathematical model showing the relationship between the application rate (*D* dose) of a given plant protection product (PPP) and the level residue of active ingredient (AI) of the substance in mature apples. This model allows the assessment of MRL values for pesticides currently approved for the protection of apple orchards and prediction of the residue level of the substances whose application is currently under consideration, when their effective doses are known, as well as any other, which may be used in apple orchards in an undefined future. Using the proposed model, we could have assessed the current MRLs. Most of them, currently found in apple orchards protection programs in Poland, as it turned out, were fixed at too high levels. We have made a comparison between level of some pesticide residues found in the mature apples within the framework of National Monitoring Programs of residue surveys carried out in Poland and the Czech Republic, and the initial level of the residue of the same substances predicted by the model. This comparison clearly has shown that the MRLs currently in force in the EU countries require verification. In the creation of a coherent system MRL an equation $R_t = R_0\, e^{-kt}$ may be helpful, especially$R_t = 1.2593 \times D \times e^{-kt}$, which takes into account the relationship between the application rate of the plant protection product and the residue level of an active ingredient in the mature apples.

REFERENCES

1. Alary J, Bescos D, Monge MC, Debrauwer L, Bories G (1995) Laboratory simulation of captan residues degradation during apple processing. Food Chem 54(2):205–211

2. Ambrus Á, Lantos J (2002) Evaluation of the studies on decline of pesticide residues. J Agric Food Chem 50(17):4846–4851

3. Ambrus Á (2000) Within and between field variability of residue data and sampling implication. Food Addit Contam 17(7):519–537

4. van Boxstael S, Habib I, Jacxsens L, De Vocht M, Baert L, Van De Perre E et al (2013) Food safety issues in fresh produce: bacterial pathogens, viruses and pesticide residues indicated as major concerns by stakeholders in the fresh produce chain. Food Control 32(1):190–197

5. Cabras P, Angioni A, Caboni P, Garau VL, Melis M, Pirisi FM et al (2000) Distribution of folpet on the grape surface after treatment. J Agric Food Chem 48(3):915–916

6. Drogué S, DeMaria F (2012) Pesticide residues and trade, the apple of discord? Food Policy 37(6):641–649

7. Gabriolotto C, Monchiero M, Negre M, Spadaro D, Gullino ML (2009) Effectiveness of control strategies against *Botrytis cinerea* in vineyard and evaluation of the residual fungicide concentration. J Environ Sci Health, Part B 44(4):389–396

8. MacLachlan DJ, Hamilton D (2010) Estimation methods for maximum residue limits for pesticides. Regul Toxicol Pharmacol 58(2):208–218

9. Malhat F, Kamel E, Saber A, Hassan E, Youssef A, Almaz M et al (2013) Residues and dissipation of kresoxim methyl in apple under field condition. Food Chem 140(1–2):371–374

10. Mouron P, Heijne B, Naef A, Strassemeyer J, Hayer F, Avilla J et al (2012) Sustainability assessment of crop protection systems: sustainOS methodology and its application for apple orchards. Agric Syst 113:1–15

11. Nowacka A, Gnusowski B, Walorczyk S, Drożdżyński D, Raczkowski M, Hołodyńska A et al (2011) Pozostałości środków ochrony roślin w płodach rolnych (rok 2010). Prog Plant Prot 50(4):1723–1738

12. Nowacka A, Gnusowski B, Walorczyk S, Drożdżyński D, Wójcik A, Raczkowski M et al (2009) Pozostałości środków ochrony roślin w płodach rolnych (rok 2008). Prog Plant Prot 49(4):1903–1917

13. Popp J, Peto K, Nagy J (2013) Pesticide productivity and food security. A Rev Agron Sustain Dev 33:243–255

14. Renwick AG (2002) Pesticide residue analysis and its relationship to hazard characterization (ADI/ARfD) and intake estimations (NEDI/NESTI). Pest Manag Sci 58(10):1073–1082

15. Reuveni M (2002) Efficacy of trifloxystrobin (Flint), a new strobilurin fungicide, in controlling powdery mildews on apple,

mango and nectarine, and rust on prune trees. Crop Prot 19(5):335–341

16. Reynolds SL, Fussel RJ, MacArthur R (2005) Investigation into validity of extrapolation in setting maximum residue levels for pesticides in crops of similar morphology. Food Addit Contam 22(1):31–38

17. Sadło S, Szpyrka E (2001) Behaviour of pyrimethanil residues on tomato plants. J Plant Prot Res 41(3):302–308

18. Sadło S (1998) Partition coefficient—its determination and significance in estimation of pesticide residue losses in the course of extraction procedure. J Plant Prot Res 38(2):179–184

19. Solecki R, Davies L, Dellarco V, Dewhurst I, van Raaij M, Tritscher A (2005) Guidance on setting of acute reference dose (ARfD) for pesticides. Food Chem Toxicol 43(11):1569–1593

20. Stěpán R, Tichá J, Hajšlová J, Kowalczuk T, Kocourek V (2005) Baby food production chain: pesticide residues in fresh apples and products. Food Addit Contam 22(12):1231–1242

21. Valverde-Garcia A, Gonzalez-Pradas E, Aguilera-des Real A (1993) Analysis of buprofezin residues in vegetables. Application to the degradation study on eggplant grown in a greenhouse. J Agric Food Chem 41(12):2319–2323

22. Walorczyk S, Drożdżyński D (2011) Development and validation of a routine multiresidue method for determining 140 pesticides in fruits and vegetables by gas chromatography/tandem quadrupole mass spectrometry. J AOAC Int 94(5):1625–1642

23. Wilson J, Otsuki T (2004) To spray or not to spray: pesticides, banana exports, and food safety. Food Policy 29(2):131–146

24. Winchester N, Rau M-L, Goetz C, Larue B, Otsuki T, Shutes K et al (2012) The impact of regulatory heterogeneity on agro-food trade. World Econ 35(8):973–993

Chapter 7

Evaluation of Chemical and Biological Seed Treatments to Control Charcoal Rot of Soybean

Sebastian Reznikov[1] • Gabriel R. Vellicce[1] • Victoria Gonza´lez[1] • Vicente de Lisi[1] • Atilio P. Castagnaro[1] • L. Daniel Ploper[1]

Instituto de Tecnología Agroindustrial del Noroeste Argentino (ITANOA) Estación Experimental Agroindustrial Obispo Colombres (EEAOC)-Consejo Nacional de Investigaciones Científicas y Técnicas (CONICET)Las Talitas Argentina

ABSTRACT

Charcoal rot caused by *Macrophomina phaseolina* (Mp) affects economically important crops around the world. In northwestern Argentina, the disease is present in seasons with dry and hot weather. Here, the efficacy of seed treatments with two biological products (*Trichodermaviride*or *Bacillus subtilis*) or one chemical treatment (thiophanate methyl + pyraclostrobin) were evaluated in the field to control Mp on two soybean cultivars in Tucumán, Argentina. In field tests, Mp was added to the soil at planting in the 2010/2011 and 2011/2012 growing seasons. Plant emergence (PE) was reduced for both cultivars: 6.7 and 56.8 % for NA8000 RG and 12.5 and 61.3 % for Munasqa RR in 2010/2011 and 2011/2012, respectively. On a 1–5 disease severity scale, the inoculated control of NA8000 RG scored 2.5 and Munasqa RR 1.8 during the 2010/2011 season, and 3.6 for NA8000

RG, and 2.7 for Munasqa RR during 2011/2012. Likewise, Mp CFU/g values in the inoculated control increased during the 2011/2012 season: from 2100 to 2366 in NA8000 RG and from 300 to 566 in Munasqa RR. Crop yield and 1000-seed weight were also reduced by fungal infection. Although the treatments counteracted PE reduction in both genotypes, disease severity was higher in NA8000 RG than in Munasqa RR, coincident with the highest CFU/g values. The 1000-seed weight of the two cultivars did not differ significantly among treatments. The highest yields were obtained in chemically treated plots, followed by those treated with *Trichoderma* or *Bacillus*.

Keywords: *Macrophomina phaseolina Glycine max Trichodermaviride Bacillussubtilis*

INTRODUCTION

Soybean [*Glycine max* (L.) Merr.] plants are susceptible to root and stem base rots caused by soil pathogens at all growth stages. One of these diseases is charcoal rot of soybean, caused by the polyphagous fungus *Macrophomina phaseolina* (Tassi) Goid (Mp). This pathogen infects a wide host range of nearly 500 species in more than 100 families around the world, including other important crops such as cotton (*Gossypium* spp.), chickpea (*Cicer arietinum*), corn (*Zea mays*), and common bean (*Phaseolus vulgaris*) (Srivastava et al. 2001). Morphologically, physiologically, genetically, and pathogenically, the fungus varies widely, enabling it to adapt to different environmental conditions and hence become widely distributed geographically (Su et al. 2001).

Soybean seedlings affected by Mp develop reddish brown lesions on the hypocotyl, which becomes ash-gray and then turns black (Wyllie 1989). The presence of small, black sclerotia in the cortical tissue confers the charcoal appearance that gives the disease its name (Mengistu et al. 2015). In addition to these symptoms and signs, mature plants develop chlorotic lesions on their leaves, which then die but remain attached to the stem, and finally the plants die prematurely (Ploper and Scandiani 2009). A combination of water stress and high temperatures favors disease development (Mihail 1992).

Charcoal rot of soybean is an economically important disease in North and South America, Asia, Australia, Africa and Europe (Wrather et al. 1997). Worldwide in 2006, yield losses caused by charcoal rot reached 4.2 % (Wrather et al. 2010). Yield loss in field experimental plots was estimated to be 30 % (Mengistu et al. 2011). In Argentina, hot, dry weather prevailed in the 2000/2001 growing season, which favored charcoal rot development on soybean. The provinces of Catamarca, Chaco, Córdoba, Entre Rios, Santa Fe, Salta, Santiago del Estero and Tucumán were affected with varying levels of yield losses, including total losses in some fields (Ploper et al. 2001).

Biological agents can provide an alternative to control certain plant diseases, especially when other methods such as chemicals are difficult to use. Cook and Baker (1983) successfully controlled *Sclerotium rolfsii*, *Rhizoctonia solani*, and *Pythium* spp. using antagonistic microorganisms such as *Trichoderma* spp., *Penicillium* spp., *Aspergillus* spp., and some bacteria. Jeyarajan et al. (1993) reported that a talc-based formulation of *Trichoderma viride*, produced on a commercial scale, was widely used by farmers to treat seeds of sesame (*Sesamum indicum*), groundnut (*Arachis hypogaea*), sunflower (*Helianthus annus*), chickpea and mung bean (*Vigna radiata*) for the biocontrol of root rot disease caused by Mp. Sankar and Jeyarajan (1996) studied *Trichoderma* spp. and *Gliocladiumvirens* as seed treatment for sesame (*Sesamum indicum*) to control Mp and concluded that treating the seeds with the formulation and storing them was more advantageous than storing the formulation and treating the seeds just before planting. Elad et al. (1986) found that *Trichoderma harzianum* inhibited the in vitro linear growth and microsclerotia production of Mp. When the biocontrol agent was applied to seeds, disease incidence was reduced in bean and melon (*Cucumis melo*) in the greenhouse and in melons and corn in the field.

To control fungal pathogens, chemical products are available for application of soil, seeds and/or foliage. Fungicide seed treatments are intended to control diseases that cause seed rot and damping-off before and after emergence. Mueller et al. (1999) used thiram, fludioxonil, and captan + pentachloronitrobenzene + thiabendazole to control *Sclerotinia sclerotiorum* in field trials and managed to reduce sclerotia formation in infected soybean seed by 98.0 %. Moreover, soybean seeds treated with thiram reduced the incidence of *Phomopsis sojae* and increased seed germination (Hepperly and Sinclair 1978). In studies on the control of *R. solani*, Martin et al. (1984) demonstrated differences in sensitivity to different fungicides among *R. solani* isolates, and Dorrance et al. (2003) assessed five fungicide seed treatments to control *R. solani* on soybean in the greenhouse. However, none of the evaluated seed treatments completely controlled the four tested isolates.

Currently, no chemicals are available to control charcoal rot in soybean. Mengistu et al. (2015) mentioned that seed treatments may be helpful if soybean seeds are infected with Mp, but there is no information on specific active ingredients effective against this pathogen.

Among fungicides tested against Mp in infected cotton seeds, carbendazim, quintozene and benomyl enhanced plant emergence and disease control (Chauhan 1986a, b, 1988; Dwivedi and Ghaube 1985). Monceren, Pencycuron, Carboxin 200, tolclofos-methyl, and Maximum AP also increased the percentage of surviving seedlings from Mp-infected seeds (Aly et al. 2001; Omar2005). However, whether any seed treatments can efficiently control Mp of soybean in northwestern Argentina is not known. Thus, we analyzed the efficacy of one chemical and two biological products in controlling charcoal rot of soybean and their effects on plant emergence, disease severity, Mp colony forming units (CFU), crop yield and 1000-seed weight in two soybean cultivars maturity group (MG) VIII inoculated with Mp in Tucumán, Argentina.

MATERIALS AND METHODS

Macrophomina phaseolina isolation

During the 2009/2010 season, plants with typical charcoal rot symptoms were collected from commercial fields in Tucumán and taken to the Phytopathology Laboratory of the Estación Experimental Agroindustrial Obispo Colombres (EEAOC) to isolate and characterize*Macrophomina phaseolina* (Mp). Infected samples were rinsed, and 0.5 cm sections were surface sterilized with 70 % ethanol for 30 s, and with a 5 % NaClO solution for 1 min. These sections were later rinsed with sterile water and then dried with sterile air. Samples were cultivated on potato dextrose agar (PDA, Difco, Sparks, Maryland, USA) plates with 0.2 % v/v lactic acid at 28 ± 2°C for 5 days, after which cultures were maintained at 4°C. Single sclerotia were removed using a stereomicroscope and transferred to a new PDA plate acidified with 2 % lactic acid. Then, after 24–48 h, pure cultures of each isolate were obtained, which were reserved for further analysis.

Field trials

The field trials were done at the Monte Redondo substation of the EEAOC research station in Cruz Alta department, Tucumán, Argentina (26°49'S, 64°51'W). On 29 December 2010 and 28 December 2011, seeds of MG VIII soybean cultivars NA8000 RG and Munasqa RR were planted by hand in four replications of randomized blocks, and each treatment comprised four 3-m-long rows, with a planting density of 23 seeds per m. For inoculum in field trials, we used sterile rice that had been inoculated with Mp isolate number 55 and incubated for 20 days at 30 °C in darkness to promote microsclerotium development. When soybean seeds were planted, 5 g per linear meter of this rice inoculum was added by hand at planting depth. Treatments were as follows: (1) uninoculated and untreated control (neither inoculation, nor chemical or biological treatments); (2) untreated control inoculated with Mp; (3) Mp added to the soil and seed treated with a mixture of strobilurin + benzimidazole (pyraclostrobin 5 % + thiophanate methyl 45 %, 100 mL/100 kg of seed); (4) Mp added to the soil and seed treated with*Trichodermaviride*. (Trichoderma SP®, *Trichoderma viride*, 700 mL/50 kg of seed, Laboratorio San Pablo, Tucumán, Argentina); and (5) Mp added to the soil and seed treated with *Bacillussubtilis* (Bacillus subtilis SP®, *Bacillus subtilis*, 500 mL/50 kg of seed, Laboratorio San Pablo, Tucumán, Argentina).

The effects of each treatment were evaluated considering plant emergence (PE) percentage, disease severity, Mp colony forming units per gram of root (CFU/g), crop yield and 1000-seed weight (W 1000). PE was calculated as the percentage of plants that emerged 7, 14, and 21 days after planting (DAP). Charcoal rot severity was recorded at stage R7 (Fehr and Caviness 1977) using the scale of Paris et al. (2006): 1 = no discoloration and no microsclerotia visible; 2 = no discoloration of vascular tissue, with very few microsclerotia visible in the pith, vascular tissue or under the epidermis; 3 = partially discolored vascular tissue, with microsclerotia partially covering the tissue; 4 = discolored vascular tissue, with numerous microsclerotia visible in the tissue under the outer epidermis, in stem and root sections; 5 = vascular tissue with numerous microsclerotia producing a dark color inside and outside of the stem and root tissue.

Values of CFU were calculated according to the protocol proposed by Mengistu et al. (2007), using 10 plants from each plot. The samples were obtained by cutting 10 cm above and below soil level, ensuring that they included root and stem tissue, then washed and rinsed three times with tap water to remove any soil, dried at room temperature and ground. From each sample, 0.005 g of ground tissue was transferred to a test tube, disinfected with 5 % NaClO for 1 min, and rinsed with sterile distilled water for 1 min, three times. Then 5 mL of sterile PDA was added at 60 °C and poured into sterile Petri dishes, which were incubated at 28 °C for 3–5 days for CFU to be recorded.

Crop yield (kg/ha) and 1000-seed weight (g) were measured at harvest.

Data for crop yield and 1000-seed-weight were analyzed for statistical differences among treatment means using an ANOVA and LSD test at $\alpha = 0.05$ in INFOSTAT software version 2008 (Balzarini et al. 2008). Data on PE, CFU and disease severity were square-root transformed ($\sqrt{x}$) before the analysis.

To evaluate whether prevailing environmental conditions during the growing seasons corresponded to those that favor charcoal rot development, average monthly temperatures and precipitation were compared for the 2010/2011 and 2011/2012 growing seasons (Table 1).

Table 1. Mean monthly temperature and rainfall in the 2010/2011 and 2011/2012 seasons in Monte Redondo (Cruz Alta, Tucumán) compared with historical average values for this site (average temperature and precipitation in 1996–2012 and 1981–2010, respectively)

	November		December		January		February		March		April		Total
	T[a]	P[b]	T[a]	P[b]	T[a]	P[b]	T[a]	P[b]	T[a]	P[b]	T[a]	P[b]	P[b]
2010/2011	23.2	50.8	25.6	159.3	25.5	208.3	23.7	162.6	22.2	109.0	20.2	46.0	736
Differences	−0.9	−41.7	+0.1	+4.9	−0.4	+21.2	−0.8	+26.2	−0.7	−29.6	+0.5	−16.2	−35.2
2011/2012	25.5	67.3	25.4	84.1	27.1	84.6	27.3	81.3	23.8	71.4	19.9	93.0	481
Differences	+1.4	−25.2	−0.1	−70.3	+1.2	−102.5	+2.8	−55.1	+0.9	−67.2	+0.2	+30.2	−290
Mean values	24.1	92.5	25.5	154.4	25.9	187.1	24.5	136.4	22.9	138.6	19.7	62.2	771

[a] *T* monthly average temperature

[b] *P* rainfall in mm

RESULTS

Using the *Macrophomina phaseolina* (Mp) isolate number 55, we were able to reproduce the disease in the field. Inoculation reduced the number of inoculated NA8000 RG control plants that had emerged (PE), for instance, at 21 DAP in the 2010/2011 season to 86.4 %, compared with 92.6 % for the noninoculated control, although without statistically significant differences (Table 2). But in the 2011/2012 season, by 21 DAP, the inoculated control had significantly lower PE values (22.4 %), than the noninoculated control (51.9 %). Similarly, in Munasqa RR by 21 DAP in the 2010/2011 season, the PE of 79.2 % for the inoculated control differed significantly from 90.5 % PE for the noninoculated control (Table 3). Moreover, by 21 DAP in 2011/2012, the inoculated control had a PE of 25.5 %, differing significantly from 65.9 % for the noninoculated control.

Table 2. NA8000 RG emergence by 7, 14 and 21 days after planting in field trials after treatment with chemicals or biological control agents in Monte Redondo (Cruz Alta, Tucumán) in 2010/2011 and 2011/2012 seasons

Treatment	2010/2011			2011/2012		
	7	14	21	7	14	21
1-Untreated and uninoculated control	89.2 a*	92.4 a	92.6 a	30.1 a	51.9 ab	51.9 a
2-Untreated control + Mp	84.6 a	86.2 b	86.4 a	16.4 ab	30.8 c	22.4 c
3-(Pyraclostrobin + thiophanate methyl) + Mp	84.8 a	92.4 a	90.2 a	22.8 ab	62.8 a	58.9 a
4-*Trichoderma viride* + Mp	84.8 a	88.6 ab	86.4 a	19.5 ab	43.5 bc	36.3 b
5-*Bacillus subtilis* + Mp	84.1 a	88.2 ab	86.4 a	13.9 b	36.1 c	26.4 bc

* Means in each column followed by the same letter are not significantly different (LSD, $\alpha = 0.05$)

Table 3. Emergence of Munasqa RR plants by 7, 14 and 21 days after planting in field trials after treatment with chemicals or biological control agents in Monte Redondo (Cruz Alta, Tucumán), 2010/2011 and 2011/2012 seasons

Treatment	2010/2011			2011/2012		
	7	14	21	7	14	21
1-Untreated and uninoculated control	90.7 a*	90.0 a	90.5 a	52.6 ab	73.6 a	65.9 a
2-Untreated control + Mp	60.2 d	82.8 b	79.2 b	32.5 c	30.8 b	25.5 b
3-(Pyraclostrobin + thiophanate methyl) + Mp	87.1 ab	92.0 a	91.1 a	56.0 a	78.6 a	61.5 a
4-*Trichoderma viride* + Mp	75.2 c	88.1 ab	88.8 a	37.8 bc	43.3 b	34.4 b
5-*Bacillus subtilis* + Mp	78.2 bc	87.1 ab	88.1 a	32.7 c	40.6 b	32.2 b

* Means in each column followed by the same letter are not significantly different (LSD, $\alpha = 0.05$)

The effects of different treatments on PE of seeds inoculated with Mp 7, 14 and 21 DAP in NA8000 RG and Munasqa RR are also shown in Tables 2 and 3, respectively. In NA8000 RG during the 2010/2011 season, only the treatment with pyraclostrobin + thiophanate methyl (PE: 92.4 %) reduced the adverse effect of inoculation with Mp and differed significantly from the inoculated control (86.2 %) at 14 DAP. PE values at 7 and 21 DAP did not differ significantly among treatments (Table 2).

The 2011/2012 season showed environmental conditions favorable for charcoal rot (Table 1). As reflected in Table 2, significant differences were observed between the Mp-inoculated and the noninoculated control of NA8000 RG at 14 and 21 DAP; the inoculated plants had lower PE values (16.4 % by 7 DAP, 30.8 % by 14 DAP and 22.4 % by 21 DAP). At 7 DAP, none of the Mp-inoculated plants treated with either chemicals or biological agents differed significantly in PE from the untreated Mp-inoculated control (Table 2). At 14 DAP, the PE for plants in treatment 3 [(pyraclostrobin + thiophanate methyl) + Mp] was 62.8 % and differed significantly from the rest of the treatments except for the untreated and uninoculated control (51.9 %). At 21 DAP, all treatments yielded statistically significant differences in PE in relation to the inoculated control (22.4 %), except treatment 5 (*B. subtilis* + Mp) (26.4 %). Treatments 1 (untreated and uninoculated control) (51.9 %) and 3 ([pyraclostrobin + thiophanate methyl] + Mp) (58.9 %) had the highest PE values with no differences between them, whereas treatment 4 (*T. viride* + Mp) (36.3 %) and treatment 5 (*B. subtilis* + Mp) (26.4 %) gave the lowest PE values. It is worth mentioning that treatment 5 did not differ from the inoculated control (22.4 %) at this stage.

PE values for Munasqa RR after the different treatments are shown in Table 3 for three dates during the 2010/2011 and 2011/2012 seasons. In 2010/2011, the PE for the inoculated and the noninoculated control differed significantly at all dates. Seven days after planting, all treatments had PE values that contrasted significantly with those of the inoculated control (60.2 %). At 14 DAP, treatments 4 (*T. viride* + Mp) (88.1 %) and 5 (*B. subtilis* + Mp) (87.1 %) did not differ statistically from the inoculated control (82.8 %), but treatment 3 ([pyraclostrobin + thiophanate methyl] + Mp) (92.0 %) did show significant differences. At 21 DAP, all treatments differed statistically from the inoculated control (79.2 %), but not among them.

For Munasqa RR in the 2011/2012 season, the PE of plants in the untreated and uninoculated control (52.6 %) and treatment 3 [(pyraclostrobin + thiophanate methyl) + Mp] (56.0 %) differed statistically from the inoculated control (32.5 %) at 7 DAP (Table 3). On the other hand, the PE of plants in treatments 4 (*T. viride* + Mp) (37.8 %) and 5 (*B. subtilis* + Mp) (32.7 %) did not differ statistically from the inoculated control (32.5 %). At 14 DAP, the PE of inoculated control (30.8 %) plants did not differ statistically from treatments 4 (*T. viride* + Mp) (43.3 %) and 5 (*B. subtilis* + Mp) (40.6 %) but differed from treatments 1 (untreated and uninoculated control) (73.6 %) and 3 ([pyraclostrobin + thiophanate methyl] + Mp) (78.6 %). At 21 DAP, the PE for treatments 4 (*T. viride* + Mp) (34.4 %) and 5 (*B. subtilis* + Mp) (32.2 %) did not differ from the inoculated control (25.5 %), but differed significantly from the PE for treatments 1 (untreated and uninoculated control) (65.9 %) and 3 ([pyraclostrobin + thiophanate methyl] + Mp) (61.5 %).

The decrease in PE values in the 2011/2012 season compared with those in the 2010/2011 season is primarily attributable to the marked differences in environmental conditions, especially with regard to precipitation from December to January, key months for root emergence and plant establishment. According to data in Table 1, in the first season, the combined precipitation of those 2 months amounted to 367.6 mm, whereas in the second season that value dropped to 168.7 mm. This low value of precipitation was accompanied by high temperatures in January 2012 (Table 1), resulting in favorable environmental conditions for charcoal rot development in the 2011/2012 seasons.

Table 4 shows data related to crop yield, 1000-seed weight and charcoal rot severity recorded in NA8000 RG in the 2010/2011 and 2011/2012 seasons. In 2010/2011, although no significant differences were found among treatments in relation to crop yield, crop yield increased with all control treatments. For instance, treatment 3 ([pyraclostrobin + thiophanate methyl] + Mp) led to 277.7 kg/ha more soybeans than the inoculated control (4492.4 kg/ha). With respect to 1000-seed weight, no significant differences were observed among treatments. Severity value in the inoculated control was 2.5, and no significant differences were found among treatments (Table 4).

In the 2011/2012 season, treatment 3 ([pyraclostrobin + thiophanate methyl] + Mp) (1345.8 kg/ha) and treatment 1 (untreated and uninoculated control) (1445.8 kg/ha) differed statistically from treatment 2 (inoculated control) (504.2 kg/ha), treatment 4 (*T. viride* + Mp) (804.2 kg/ha) and treatment 5 (*B. subtilis* + Mp) (741.7 kg/ha) (Table 4). Treatment 3 ([pyraclostrobin + thiophanate

Table 4. Mean crop yield (kg/ha), 1000-seed weight (W 1000 in g) and charcoal rot severity (Sev) on a scale from 1 to 5 in NA8000 RG in field trials after treatment with chemicals or biological control agents in Monte Redondo (Cruz Alta, Tucumán), 2010/2011 and 2011/2012 seasons

Treatment	2010/2011			2011/2012		
	Yield	W 1000	Sev	Yield	W 1000	Sev
1-Untreated and uninoculated control	4515.1 a*	145.6 a	1.0 a	1445.8 a	170.0 b	1.3 c
2-Untreated control + Mp	4492.4 a	148.6 a	2.5 a	504.2 b	169.7 b	3.6 a
3-(Pyraclostrobin + thiophanate methyl) + Mp	4770.1 a	145.7 a	2.0 a	1345.8 a	179.9 ab	2.6 b
4-*Trichoderma viride* + Mp	4687.3 a	146.0 a	1.5 a	804.2 b	186.7 a	2.7 b
5-*Bacillus subtilis* + Mp	4631.2 a	144.6 a	1.0 a	741.7 b	181.9 ab	3.1 ab

* Means in each column followed by the same letter are not significantly different (LSD, $\alpha = 0.05$)

methyl] + Mp) resulted in 841.6 kg/ha more in yield than the inoculated control. As for 1000-seed weight, treatment 4 (*T. viride* + Mp) (186.7 g) differed statistically with respect to the untreated and uninoculated (170.0 g) and the inoculated (169.7 g) controls, but not with respect to treatments 3 [(pyraclostrobin + thiophanate methyl) + Mp] (179.9 g) and 5 (*B. subtilis* + Mp) (181.9 g). The inoculated control showed high charcoal rot severity values (3.6) and differed statistically from the non-inoculated control (1.3), treatment 3 ([pyraclostrobin + thiophanate methyl] + Mp) (2.6) and treatment 4 (*T. viride* + Mp) (2.7), but not from treatment 5 (*B. subtilis* + Mp) (3.1) (Table 4). Again, the higher charcoal rot values in 2011/2012 compared with 2010/2011 can be accounted for by the favorable environment conditions that prevailed during that season (Table 1).

In Munasqa RR, yield of all treatments differed significantly from the fungus-inoculated control (3456.1 kg/ha) in the 2010/2011 season (Table 5). Treatment 3 ([pyraclostrobin + thiophanate methyl] + Mp) yielded 658.3 kg/ha more than the inoculated control, while treatments 4 (*T.viride* + Mp) and 5 (*B.subtilis* + Mp) showed increments of 635.6 kg/ha and 634.3 kg/ha, respectively. No significant differences were observed for 1000-seed weight among different treatments and the inoculated and uninoculated controls. With regard to disease severity, no significant differences were found among treatments and the inoculated control, which showed a rating of 1.8.

Table 5. Mean yield (kg/ha), 1000-seed weight (W 1000 in g) and charcoal rot severity (Sev) on a scale from 1 to 5 in Munasqa RR cultivar in field trials after treatment with chemicals or biological control agents in Monte Redondo (Cruz Alta, Tucumán), 2010/2011 and 2011/2012 seasons

Treatment	2010/2011			2011/2012		
	Yield	W 1000	Sev	Yield	W 1000	Sev
1-Untreated and uninoculated control	4166.5 a*	137.3 a	1.0 a	1866.7 a	163.2 ab	1.2 b
2-Untreated control + Mp	3456.1 b	139.4 a	1.8 a	870.8 c	169.4 a	2.7 a
3-(Pyraclostrobin + thiophanate methyl) + Mp	4114.4 a	140.8 a	1.5 a	1945.8 a	151.4 b	2.5 a
4-*Trichoderma viride* + Mp	4091.7 a	140.5 a	1.0 a	1491.7 b	160.8 ab	2.5 a
5-*Bacillus subtilis* + Mp	4090.4 a	140.0 a	1.5 a	1304.2 b	158.8 ab	2.5 a

* Means in each column followed by the same letter are not significantly different (LSD, $\alpha = 0.05$)

In the 2011/2012 season, yields for treatment 3 [(pyraclostrobin + thiophanate methyl) + Mp] (1945.8 kg/ha) and the noninoculated control (1866.7 kg/ha) differed statistically from that of the Mp-inoculated control (870.8 kg/ha) and treatments 4 (*T. viride* + Mp) (1491.7 kg/ha) and 5 (*B. subtilis* + Mp) (1304.2 kg/ha). Treatment 3 ([pyraclostrobin + thiophanate methyl] + Mp) yielded 1075.0 kg/ha more than the inoculated control and 79.1 kg/ha more than the noninoculated control. With regards to 1000-seed weight, treatment 3 ([pyraclostrobin + thiophanate methyl] + Mp) (151.4 g) differed statistically from the inoculated control (169.4 g). The Mp-inoculated control showed the highest charcoal rot severity values (2.7), but differed statistically only from the noninoculated control (1.2). The fact that higher charcoal rot severity values (2.7) were observed in the 2011/2012 season than in the 2010/2011 season (1.8) can also be accounted for by environmental conditions (Table 1).

With respect to CFU/g of root, the Mp-inoculated control presented higher CFU/g values than in the other treatments for both cultivars and seasons. NA8000 RG presented 2100 CFU/g in 2010/2011 and differed significantly from all the other treatments (Fig. 1). In 2011/2012, Mp-inoculated control (2366 CFU/g) differed significantly from treatments 1 (untreated and uninoculated control) (166 CFU/g), 3 ([pyraclostrobin + thiophanate methyl] + Mp) (834 CFU/g) and 4 (*T. viride* + Mp) (934 CFU/g), but not from treatment 5 (*B. subtilis* + Mp) (2100 CFU/g). On the other hand, in 2010/2011 Munasqa RR inoculated control showed 300 CFU/g and differed statistically from treatments 1 (untreated and uninoculated control) (50 CFU/g) and 4

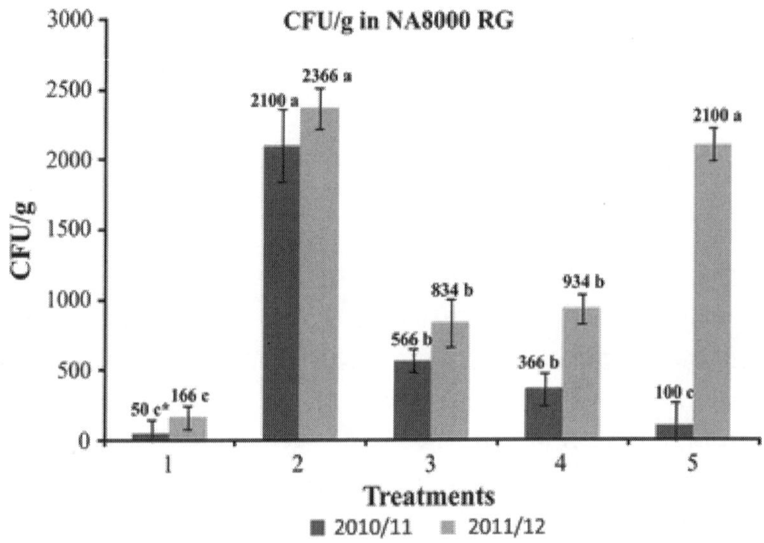

Figure 1. CFU/g of *Macrophomina phaseolina* in NA8000 RG during the 2010/2011 and 2011/2012 seasons in field trials with different seed treatments. Treatment *1* untreated and uninoculated control, *2*untreated control + Mp, *3* (pyraclostrobin + thiophanate methyl) + Mp, *4Trichodermaviride* + Mp and *5Bacillussubtilis* + Mp. * Means in each season followed by the same letter are not significantly different (LSD, $\alpha = 0.05$)

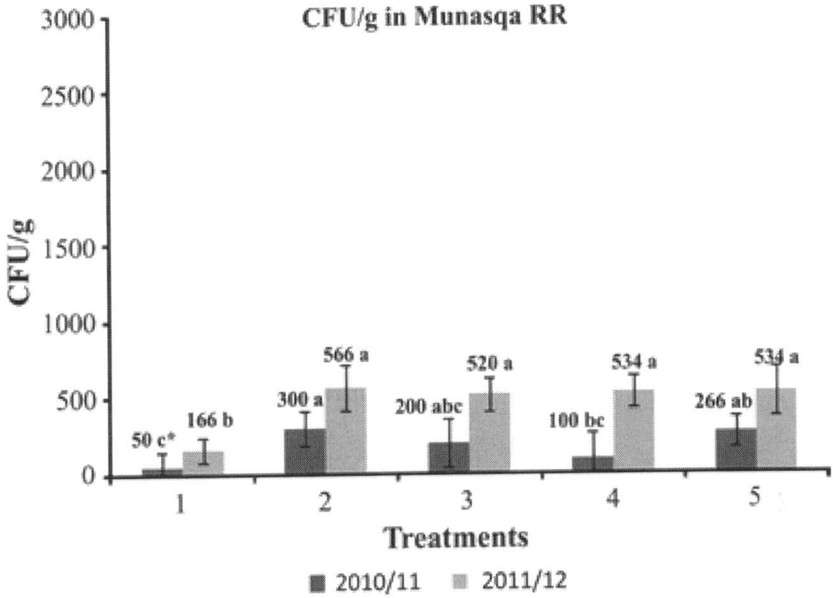

Figure 2. CFU/g of *Macrophomina phaseolina* in Munasqa RR during the 2010/2011 and 2011/2012 seasons in field trials with different seed treatments. Treatment *1* untreated and uninoculated control, *2*untreated control + Mp, *3* (pyraclostrobin + thiophanate methyl) + Mp, *4Trichodermaviride* + Mp and *5Bacillussubtilis* + Mp. * Means in each season followed by the same letter are not significantly different (LSD, α = 0.05)

(*T. viride* + Mp) (100 CFU/g) but did not from treatments 3 ([pyraclostrobin + thiophanate methyl] + Mp) (200 CFU/g) and 5 (*B. subtilis* + Mp) (266 CFU/g) (Fig. 2). In 2011/2012 season, Munasqa RR inoculated control was 566 CFU/g and differed statistically only from the noninoculated control (166 CFU/g). All treatments had lower CFU/g values than the inoculated control in both cultivars and seasons.

DISCUSSION

Field inoculation of soybean with *Macrophomina phaseolina* (Mp) affected all variables studied, as compared with the noninoculated control in NA8000 RG and Munasqa RR. In the 2010/2011 season, Mp inoculation resulted in a PE reduction of 6.7 % and 12.5 % in NA8000 RG and Munasqa RR, respectively. In the 2011/2012 season, the PE reduction was even more conspicuous and amounted to 56.8 % in NA8000 RG and 61.3 % in Munasqa RR. These results are in accordance with Smith and Wyllie (1999), who reported high seedling mortality from charcoal rot in the tropics.

Disease severity values recorded at R7 in the inoculated control were also higher in the 2011/2012 season and in both cultivars: 3.6 for NA8000 RG and 2.7 for Munasqa RR, compared with 2.5 and 1.8, respectively, as recorded in 2010/2011. In addition, values of CFU/g root in the fungus-inoculated control reached 2100 in 2010/2011 and 2366 in 2011/2012 in NA8000 RG (Fig. 1). In Munasqa RR, these values were 300 CFU/g in 2010/2011 and 566 CFU/g in 2011/2012 (Fig. 2).

The higher PE reduction, disease severity and CFU/g values observed in the 2011/2012 season were likely the consequences of climatic conditions, with high temperatures and low rainfall levels (Table 1). Similarly, Mengistu et al. (2013) greater symptoms of charcoal rot in plants stressed by high heat and drought, especially when drought occurred during reproductive growth.

Genotypes also differ in their reaction to charcoal rot. Mengistu et al. (2013) identified sources of charcoal rot resistance in 628 accessions from soybean germoplasm by field screening in five environments. In this study, we found that Mp on Munasqa RR was less severe and developed fewer CFU/g than on NA8000 RG in both seasons.

All treatments evaluated in this research were effective in counteracting PE reduction caused by the pathogen in both seasons and cultivars. In 2011/2012 the fungicide treatment led to higher PE values than the biological treatments in both cultivars. In NA8000 RG, compared with the inoculated control, the fungicide enhanced PE by 70.3 %, *T.viride* by 26.8 % and *B.subtilis* by 7.7 %. As for Munasqa RR, the fungicide increased PE by 54.6 %, *T.viride* by 13.5 % and*B.subtilis* by 6.1 % with respect to the inoculated control. The lower effectiveness showed by the biological treatments could be due to the fact that weather conditions (Table 1) favoring Mp infection (high temperature and drought) could have affected these microorganisms negatively.

Biological and chemical seed treatments have been shown to be effective to control Mp on various crops. For cotton, several fungicides were tested as seed treatments to determine their efficacy in controlling Mp infection. Treatments with carbendazim, quintozene and benomyl enhanced plant emergence and disease control (Chauhan 1986a, b, 1988; Dwivedi and Ghaube1985). Omar (2005) found that Monceren® 250 FS and tolclofos-methyl were the best performing fungicides in controlling Mp on cotton in the greenhouse. Pineda and Gonnella (1988) also found a reduction in the percentage of dead sesame plants affected by Mp when seeds were treated with biological antagonists such as *Trichoderma* sp. and *Aspergillus* sp.

In NA8000 RG and Munasqa RR, treatments had a similar performance trend: the highest crop yield values were obtained with the pyraclostrobin + thiophanate methyl mixture, followed by those obtained with *T.viride* and *B.subtilis.* Similar yield increases were also observed by Pineda and Avila (1988): an increment of 100 kg was obtained when applying chemical seed treatments (propineb and dicarboximide) in sesame to control Mp, as compared with yield values obtained with the untreated control.

Mp did infect Munasqa RR and NA8000 RG, but severity levels and CFU/g were higher in NA8000 RG in both seasons. The chemical and biological treatments resulted in lower disease severity and CFU/g values than in the inoculated control. These results agree with those reported by Ramezani (2008), who found lower disease incidence in eggplants (*Persea americana* L.) treated with different *Trichoderma* sp. strains.

Results from this study of chemical and biological seed treatments of soybean to control charcoal rot will be useful for farmers in Tucumán to develop more efficient management strategies for this important disease. Additional research should be extended to other regions of Argentina and neighboring soybean-producing countries in South America to further assess this plant–pathogen interaction under different environmental conditions.

REFERENCES

1. Aly AA, El-Shazly AMM, Youssef RM, Omar MR (2001) Chemical and biological control of charcoal rot of cotton caused by *Macrophomina phaseolina*. J Agric Sci Mansoura Univ 26:7661–7674

2. Balzarini MG, Gonzalez L, Tablada M, Casanoves F, Di Rienzo JA, Robledo CW (2008) Infostat. Manual del usuario, Editorial Brujas, Córdoba, Argentina

3. Chauhan MS (1986a) Comparative efficacy of fungicides for the control of seedling diseases of cotton due to *Rhizoctonia* spp. Ind Mycol Pl Pathol 16:335–337

4. Chauhan MS (1986b) Systemic and non-systemic fungicides against root rot of cotton in Haryana. Ind Mycol Pl Pathol 16:226–227

5. Chauhan MS (1988) Relative efficiency of different methods for the control of seedling disease of cotton by *Rhizoctonia bataticola*. Ind Mycol Pl Pathol 18:25–30

6. Cook RJ, Baker KF (1983) The nature and practice of biological control of plant pathogens. American Phytopathological Society, St. Paul

7. Dorrance AE, Kleinhenz MD, McClure SA, Tuttle NT (2003) Temperature, moisture, and seed treatment effects on *Rhizoctonia solani* root rot of soybean. Plant Dis 87:533–538

8. Dwivedi TS, Ghaube HS (1985) Effect of fungicides on the emergence and infection of cotton seedlings by *Macrophomina phaseolina* (Tassi) Goid. Ind Mycol Pl Pathol 15:295–296

9. Elad Y, Zvieli Y, Chet I (1986) Biological control of *Macrophomina phaseolina* (Tassi) Goid by*Trichoderma harzianum*. Crop Protect 5:288–292

10. Fehr WR, Caviness CE (1977) Stages of soybean development. Special report no. 80. Coop. Ext. Ser., Iowa Agric And Home Econ Exp Stn, Iowa State Univ, Ames

11. Hepperly PR, Sinclair JB (1978) Quality losses in *Phomopsis*-infected soybean seeds. Phytopathology 68:1684–1687

12. Jeyarajan R, Ramakrishman G, Dinakaran D, Sridhar R (1993) Development of product of*Trichoderma viride* and *Bacillus subtilis* for biocontrol of root rot disease. In: Dwivedi (ed) Biotechnology in India. Bioved Research Society, Allahabad, India

13. Martin SB, Lucas LT, Campbell CL (1984) Comparative sensitivity of *Rhizoctonia solani* and*Rhizoctonia*-like fungi to selected fungicides in vitro. Phytopathology 74:778–781

14. Mengistu A, Ray JD, Smith JR, Paris RL (2007) Charcoal rot disease assessment of soybean genotypes using a colony-forming unit index. Crop Sci 47:2453–2461

15. Mengistu A, Smith JR, Ray JD, Bellaloui N (2011) Seasonal progress of charcoal rot and its impact on soybean productivity. Plant Dis 95:1159–1166

16. Mengistu A, Bond J, Nelson R, Rupe J, Shannon G, Arelli P, Wrather A (2013) Identification of soybean accessions resistant to *Macrophomina phaseolina* by field screening and laboratory validation. Online. Plant Health Progress

17. Mengistu A, Wrather A, Rupe JC (2015) Charcoal rot. In: Hartman GL, Rupe JC, Sikora EF, Domier LL, Davies JA, Steffey KL (eds) Compendium of soybean diseases. APS Press, St. Paul, pp 67–69

18. Mihail JD (1992) *Macrophomina*. In: Rush CM (ed) Singleton LL Mihail JD. Methods for research on soilborne phytopathogenic fungi. APS Press, St. Paul, pp 134–136

19. Mueller DS, Hartman GL, Pedersen WL (1999) Development of sclerotia and apothecia of*Sclerotinia sclerotiorum* from infected soybean seed and its control by fungicide seed treatment. Plant Dis 83:1113–1115

20. Omar MR (2005) Pathological and biochemical studies on *Macrophomina phaseolina*pathogenic on cotton. Ph.D. dissertation. Suez Canal University, Ismailia, Egypt

21. Paris RL, Mengistu A, Tyler JM, Smith JR (2006) Registration of soybean germplasm line DT97-4290 with moderate resistance to charcoal rot. Crop Sci 46:2324–2325

22. Pineda JB, Avila JM (1988) Alternativas para el control de *Macrophomina phaseolina* y*Fusarium oxysporum* patógenos del ajonjolí (*Sesamum indicum* L.). Agron Trop 38:79–84

23. Pineda JB, Gonnella ER (1988) Evaluación del control biológico de *Macrophomina phaseolina*en ajonjolí (*Sesamum indicum* L.). Agron Trop 38:43–48

24. Ploper LD, Scandiani MM (2009) Visión general de las enfermedades radiculares de la soja en Argentina. Proceedings V Congresso Brasileiro de Soja e Mercosoja 2009 [CD-ROM]. Goiânia, Goias, pp 1–3

25. Ploper LD, González V, de Ramallo V, Gálvez R, Devani M (2001) Presencia de la podredumbre carbonosa del tallo de la soja en el centro y noroeste argentino. Avance Agroind 22:30–34

26. Ramezani H (2008) Biological control of root-rot of eggplant caused by *Macrophomina phaseolina*. Am Eurasian J Agric Environ Sci 4:218–220

27. Sankar P, Jeyarajan R (1996) Seed treatment formulation of *Trichoderma* and *Gliocladium* for biological control of *Macrophomina phaseolina* in sesamum. Indian Phytopathology 49:148–151

28. Smith GS, Wyllie TD (1999) Charcoal rot. In: Hartman GL, Sinclair JB, Rupe JC (eds) Compendium of soybean diseases, 4th edn. APS Press, St. Paul, pp 29–31

29. Srivastava AK, Singh T, Jana TK, Arora DK (2001) Microbial colonization of *Macrophomina phaseolina* and suppression of charcoal rot of chickpea. In: A. Sinha (ed) Microbes and plants. Vedams eBooks (P) Ltd., New Delhi, India pp 269–319

30. Su G, Suh SO, Schneider RW, Russin JS (2001) Host specialization in the charcoal rot fungus,*Macrophomina phaseolina*. Phytopathology 91:120–126

31. Wrather JA, Anderson TR, Arsyad DM, Gai J, Ploper LD, Porta-Puglia A, Ram HH, Yorinori JT (1997) Soybean disease loss

estimates for the top ten soybean producing countries in 1994. Plant Dis 81:107–110

32. Wrather JA, Shannon G, Balardin R, Carregal L, Escobar R, Gupta GK, Ma Z, Morel W, Ploper D, Tenuta A (2010) Effect of diseases on soybean yield in the top eight producing countries in 2006. Online. Plant Health Progress

33. Wyllie TD (1989) Charcoal rot. In: Sinclair JB, Backman PA (eds) Compendium of soybean diseases, 3rd edn. APS Press, St. Paul, pp 30–33

Chapter 8

The Effect of Plant Extracts as Seed Treatments to Control Bacterial Leaf Spot of Tomato in Tanzania

Ernest R. Mbega[1,2] • Carmen N. Mortensen[1] • Robert B. Mabagala[2] • Ednar G. Wulff[1]

[1]*Department of Agriculture and Ecology, Faculty of Life Sciences, Danish Seed Health Centre for Developing Countries University of Copenhagen Copenhagen Denmark*

[2]*Department of Crop Science and Production, African Seed Health Centre Sokoine University of Agriculture Morogoro Tanzania*

ABSTRACT

Bacterial leaf spot (BLS) caused by seed-borne xanthomonads is a serious disease of tomato (*Solanum lycopersicum* L.), causing significant losses in both yield and quality. To identify more effective control measures, we evaluated crude extracts from 84 plant species in in vitro and in planta assays for antibacterial activity against BLS of tomato. In the in vitro assays, 20.2 % of the tested plant extracts totally inhibited growth of bacteria when seed washings from treated seeds were plated on nutrient agar medium. In the in planta assays, 17.8 % of the tested plant extracts reduced BLS incidence by 100 % in tomato seedlings. The most effective seed treatments were obtained with extracts from *Aloe vera, Betula pendula, Coffea arabica,Glycyrrhiza uralensis, Juniperus communis, Ocimum basilicum, Quercus robur, Rheum palmatum, Rosmarinus officinalis, Ruta graveolens, Sinapis alba, Yucca schidigera* and *Salvia officinalis*. Seed treatment of tomato with these extracts completely inhibited *Xanthomonas perforans* in both in vitro and in planta assays. Extracts from *A. vera, C. arabica* and *Y. schidigera* were tested three times using tomato seeds of cultivars Tanya, Cal-J and Moneymaker in Tanzania. Treatment of tomato seeds with these extracts had a positive effect on the number of normal seeds and had no effect on seedling vigor, height and weight. These results indicate that plant extracts from *A. vera, C. arabica* and *Y. schidigera* are potential candidates for seed treatment against seed-borne xanthomonads of tomato in Tanzania.

Keywords: Plant extracts Antibacterial activity *Xanthomonas perforans* Tomato Seed treatment

INTRODUCTION

Bacterial leaf spot (BLS) incited by *Xanthomonas euvesicatoria, X. vesicatoria, X. perforans*and *X. gardneri* (Anonymous 2006; Jones et al. 2004) is a serious disease of tomato (*Solanum lycopersicum* L.) that occurs worldwide in regions with warm and humid climates (Jones et al.2000; Stall et al. 1994; Tamir-Ariel et al. 2007). The BLS xanthomonads also affect pepper (*Capsicum* spp.), reducing both fruit yield and quality. Recently, *X. campestris* pv. *raphani*(Punina et al. 2009) and *X. arboricola* (Myung et al. 2010; E. R. Mbega et al., unpublished data) were also reported as pathogens of tomato. The BLS xanthomonads can survive in seeds, plant debris and volunteer plants (Kaaya et al. 2003). Seed infection by BLS pathogens of about 40.7 % has been reported in Tanzania, causing yield losses up to 45 % (Black et al. 2001; Kaaya et al. 2003).

Management of BLS is limited to foliar applications of copper-based compounds. However, presence of strains of BLS pathogens with a high degree of tolerance to copper (Carrillo-Fasio et al. 2001; Gitaitis et al. 1992; Gore and O'Garro 1999; Lee and Cho 1996; Martin et al. 2004; Scheck et al. 1996; Shenge et al. 2007) and the considerable number of *Xanthomonas* species and races causing BLS symptoms in tomato and pepper (Jones et al. 2004) have made the control of the disease difficult.

Plants synthesize a number of compounds with antibiotic and antimicrobial properties (Amadioha 2003; Kumar and Parmar 1996; Opara and Wokocha 2008; Prakash and Rao 1997). Such compounds can therefore, be exploited as an alternative approach to manage or control BLS. Some of the known advantages of using pesticides of plant origin to control plant diseases include low mammalian toxicity, minimum health hazards and environmental pollution, and low risks of development of resistance by pathogens (Amadioha 2003; Kumar and Parmar 1996; Prakash and Rao 1997). Biopesticides may also be less expensive, easily available (because of their natural occurrence) and depending on the concentration, they may have no effects on seed viability, plant growth and food quality (Opara and Wokocha 2008).

In the present study, we applied crude extracts from a large number of plant species to tomato seeds to evaluate their potential to control BLS. The effect of the most promising extracts on seed germination and tomato seedling growth was also investigated.

MATERIALS AND METHODS

Seed samples

Tomato (*S. lycopersicum* L.) seeds of cultivars Cal-J, Tanya and Moneymaker, collected from tomato growers in Tanzania, were tested for infection as described by International Seed Federation (2007).

Seed samples that were free of infection by *Xanthomonas* spp. were used in the experiments. One thousand seeds per cultivar were surface-disinfested in 70 % ethanol for 1 min, then in 1 % sodium hypochlorite for 3 min and rinsed three times in sterile distilled water. The seeds were then transferred to Petri dishes containing sterile filter papers and allowed to air-dry overnight in a laminar flow chamber and stored at 4 °C until used.

Seed inoculation with bacterial suspensions

Surface-disinfested seeds of tomato cultivar Tanya were inoculated with *X. perforans* strain NCPPB 4321 based on its ability to cause severe BLS symptoms on various tomato cultivars, including those grown in Tanzania. Inoculum was prepared from 24-h-old bacterial cultures grown on nutrient agar (NA) medium (meat extract 3 g, Bacto peptone 5 g, Bacto agar 20 g, distilled water 1000 mL) at 28 °C. Bacterial cultures were flooded with 10 mL of sterile distilled water and gently scraped with a flamed Drigalski spatula. The inoculum suspension was then homogenized using a vortex mixer and suspended in sterile distilled water to obtain a ca. 10^8 CFU/mL (OD_{600} = 0.01) (NanoDrop, Thermo Fisher Scientific, Beverly, MA, USA). One thousand seeds of tomato were vacuum-infiltrated for 30 min with 10 mL of the bacterial suspension, and seeds were air-dried in the laminar air flow chamber at 4 °C until used.

Seed treatment with plant extracts

Plant for extractions were obtained from companies in Denmark and farmers in Tanzania as shown in Table 1. Two grams of plant material were suspended in 20 mL sterile distilled water in a 50 mL conical flask to obtain a 10 % (w/v) concentration. The conical flasks with the suspensions were briefly heated on a hot electric plate until boiling

Table 1. Origin of plant extracts and control of bacterial leaf spot (BLS) of tomato caused by *Xanthomonas perforans* after seed treatment with extracts or control treatments

Common name	Scientific name	Family	Part tested	Origin	CFU/mL	BLS-RI (%)[a]
Aloe	*Aloe vera* L.	Aloaceae	Stem	NN, DK	0.0×10^0	100
Silver birch	*Betula pendula* Roth.	Betulaceae	Leaf	NDC, DK	0.0×10^0	100
Coffee	*Coffea arabica* L.	Rubiaceae	Seed[b]	Tanzania	0.0×10^0	100
Licorice	*Glycyrrhiza uralensis* L.	Fabaceae	Stem/leaf	NDC, DK	0.0×10^0	100
Juniper	*Juniperus communis* L.	Cupressaceae	Stem	NDC, DK	0.0×10^0	100
Basil	*Ocimum basilicum* L.	Lamiaceae	Leaf	NDC, DK	0.0×10^0	100
Oak	*Quercus robur* L.	Fagaceae	Bark	NN, DK	0.0×10^0	100
Rhubarb	*Rheum palmatum* L.	Polygonaceae	Stem	NN, DK	0.0×10^0	100
Rosemary	*Rosmarinus officinalis* L.	Lamiaceae	Stem	NN, DK	0.0×10^0	100
Rue	*Ruta graveolens* L.	Rutaceae	Stem	NDC, DK	0.0×10^0	100
White mustard	*Sinapis alba* L.	Brassicaceae	Root	NDC, DK	0.0×10^0	100
Mojave yucca	*Yucca schidigera* L.	Agavaceae	Stem	NDC, DK	0.0×10^0	100
Sage	*Salvia officinalis* L.	Lamiaceae	Stem	NDC, DK	0.0×10^0	100
Nor-grape 80	*Vitis vinifera* L.	Vitaceae	Stem/leaf	NDC, DK	0.0×10^0	94
Punica	*Punica granatum* L.	Punicaceae	Stem	NDC, DK	0.0×10^0	86.7
Olive	*Olea europaea* L.	Oleaceae	Leaf	NDC, DK	0.0×10^0	81.2
Clove	*Caryophyllus aromaticus* L.	Myrtaceae	Twig	NDC, DK	0.0×10^0	75
White willow	*Salix alba* L.	Salicaceae	Bark	NN, DK	2.0×10^0	100
Grape vine	*Vitis vinifera* L.	Vitaceae	Leaf	NN, DK	1.0×10^0	79.8
Quinoa	*Chenopodium quinoa* Wild.	Chenopodiaceae	Stem	NN, DK	1.8×10^3	75
Soapwort	*Saponaria officinalis* L.	Caryophyllaceae	Stem/leaf	NN, DK	2.2×10^1	100
Ginkgo	*Ginkgo biloba* L.	Ginkgoaceae	Leaf	NDC, DK	3.3×10^4	94
Daisy	*Bellis perennis* L.	Asteraceae	Flower	NDC, DK	4.0×10^4	90.1
Inula	*Inula helenium* L.	Asteraceae	Stem	NDC, DK	1.0×10^1	86.7
Rosemary	*Rosmarinus officinalis* L.	Lamiaceae	Leaf	NN, DK	2.5×10^3	81.6
Tea	*Camellia sinensis* L.	Theaceae	Leaf	NDC, DK	1.0×10^1	81.2
Bilberry	*Vaccinium myrtillus* L.	Vacciniaceae	Leaf	NN, DK	1.0×10^1	81.2
Celery	*Apium graveolens* L.	Apiaceae	Root	NDC, DK	1.0×10^1	80
Ginseng	*Panax ginseng* C.A. Meyer.	Araliaceae	Stem/leaf	NDC, DK	1.0×10^1	80
Burnet-saxifrage	*Pimpinella saxifraga* ssp. *nigra* (Mill)	Apiaceae	Stem	NN, DK	1.0×10^6	80
Thyme	*Thymus vulgaris* L.	Lamiaceae	Leaf/stem	NDC, DK	3.9×10^3	80
Maypop	*Passiflora incarnata* L.	Passifloraceae	Flower	NN, DK	6.0×10^4	79.8
Red pepper	*Capsicum frutescens* L.	Solanaceae	Fruit	NN, DK	1.8×10^5	73.4
Chamomile	*Matricaria chamomilla* L.	Asteraceae	Flower	NN, DK	2.0×10^1	73.3
Neem	*Azadirachta indica* L.	Meliaceae	Seed	Tanzania	7.0×10^2	68.7
Quinine	*Cinchona pubescens* Vahl.	Rubiaceae	Bark	NDC, DK	9.0×10^1	66.7
High mallow	*Malva sylvestris* L.	Malvaceae	Leaf	NN, DK	9.8×10^3	66.6
Cowslip	*Primula veris* L.	Primulaceae	Stem	NN, DK	6.0×10^1	60
Sisal	*Agave sisalana* Perrine.	Agavaceae	Root	Tanzania	$>1.0 \times 10^7$	59.9
Agapanthus	*Agapanthus* sp.	Alliaceae	Leaf	NDC, DK	1.9×10^5	53.2
Sisal	*Agave sisalana* Perrine.	Agavaceae	Leaf	Tanzania	2.8×10^5	53.2
Couch grass	*Agropyron repens* (L.) Beauv.	Poaceae	Stem	NDC, DK	1.1×10^3	53.1
Elm	*Ulmus campestris* L.	Ulmaceae	Bark	NN, DK	5.5×10^2	49.5
Sisal	*Agave sisalana* Perrine.	Agavaceae	Stem	Tanzania	1.9×10^5	46.8
Horsetail	*Equisetum arvense* L.	Equisateaceae	Leaf	ND, DK	1.6×10^5	40
Black poplar	*Populus nigra* L.	Salicaceae	Twig	NN, DK	6.0×10^0	40
Elderberry	*Sambucus nigra* L.	Caprifoliaceae	Leaf	ND, DK	6.0×10^0	40
Turmeric	*Curcuma longa* L.	Zingiberaceae	Stem	ND, DK	$>1.0 \times 10^7$	40

Table 1 continued

Common name	Scientific name	Family	Part tested	Origin	CFU/mL	BLS-RI (%)[a]
Adam's needle	*Yucca filamentosa* L.	Agavaceae	Stem	NN, DK	2.4×10^4	37.5
Onion	*Allium cepa* L.	Liliaceae	Bud	ND, DK	2.0×10^6	33.5
Eucalyptus	*Eucalyptus globulus* L.	Myrtaceae	Leaf	Tanzania	3.7×10^3	33.5
Avocado	*Persea americana* L.	Lauraceae	Fruit peels	Tanzania	$>1.0 \times 10^7$	33.5
Buckthorn	*Frangula alnus* Mill.	Rhamnaceae	Bark	ND, DK	2.0×10^1	29.9
Plantain	*Plantago major* L.	Plantaginaceae	Leaf	ND, DK	6.0×10^5	29.9
Southernwood	*Artemisia abrotanum* L.	Asteraceae	Leaves	ND, DK	5.8×10^6	26.8
Lemongrass	*Cymbopogon citratus* L.	Poaceae	Leaves	ND, DK	9.8×10^5	26.8
Eucalyptus	*Eucalyptus globulus* L.	Myrtaceae	Seeds	Tanzania	$>1.0 \times 10^7$	26.8
White clover	*Trifolium repens* L.	Fabaceae	Flower	NN, DK	2.6×10^5	26.8
Couch grass	*Agropyron repens* L.	Poaceae	Leaf	ND, DK	1.9×10^5	26.7
Silver birch	*Betula pendula* Roth.	Betulaceae	Bark	ND, DK	5.0×10^3	26.7
Scots pine	*Pinus sylvestris* L.	Pinaceae	Leaf	ND, DK	6.0×10^3	26.7
Red clover	*Trifolium pratense* L.	Fabaceae	Flower	NN, DK	6.8×10^3	26.7
Ginger	*Zingiber officinale* Rosc.	Zingiberaceae	Bud	NN, DK	7.0×10^2	26.7
Yucca	*Yucca* sp.	Agavaceae	Stem/leaf	NN, DK	2.6×10^3	25
Adam's needle	*Yucca filamentosa* L.	Agavaceae	Leaf	NN, DK	2.4×10^5	25
Oak	*Quercus robur* L.	Fagaceae	Leaf	NN, DK	7.0×10^1	18.7
Soapbark	*Quillaja saponaria* Molina.	Quillajaceae	Bark	NN, DK	2.6×10^3	18.7
Chicory	*Cichorium intybus* L.	Asteraceae	Stem	ND, DK	1.0×10^5	13.5
Water mint	*Mentha aquatica* L.	Lamiaceae	Leaf	ND, DK	4.0×10^5	13.5
Irish moss	*Chondrus crispus* L.	Gigartinaceae	Leaf	ND, DK	3.5×10^5	13.3
Tansy	*Tanacetum vulgare* L.	Asteraceae	Stem/leaf	ND, DK	1.3×10^4	6.2
Neem	*Azadirachta indica* L.	Meliaceae	Leaf	Tanzania	5.0×10^5	0
Bladderwrack	*Fucus vesiculosus* L.	Fucaceae	Flower	NN, DK	6.9×10^4	0
Coral plant	*Jatropha* sp.	Euphorbiaceae	Seed	Tanzania	2.2×10^5	0
Coral plant	*Jatropha* sp.	Euphorbiaceae	Leaf	Tanzania	1.2×10^5	0
Lantana	*Lantana camara* L.	Verbenaceae	Leaf	Tanzania	5.0×10^4	0
Spiny restharrow	*Ononis spinosa* L.	Fabaceae	Leaf	ND, DK	6.0×10^5	0
Silverweed	*Pontentilla anserine* L.	Rosaceae	Leaf	ND, DK	9.6×10^3	0
Quassia	*Quassia* sp.	Simaroubaceae	Bark	NN, DK	1.3×10^5	0
Nettle	*Urtica* sp.	Urticaceae	Leaf	ND, DK	1.4×10^5	0
Mistletoe	*Viscum album* L.	Loranthaceae	Leaf/stem	ND, DK	4.6×10^5	0
Cardamon	*Ellettaria cardamomum* L.	Zingiberaceae		ND, DK	$>1.0 \times 10^7$	−10
Alfalfa	*Medicago sativa* ssp. *sativa* L.	Fabaceae	Seed	ND, DK	$>1.0 \times 10^7$	−10
Fenugreek	*Trigonella foenum-graecum* L.	Fabaceae	Seed	ND, DK	1.3×10^6	−20
Controls						
Sterile distilled water	–	–	–	–	$>1.0 \times 10^7$	0
Copper sulphate	–	–	–	–	0.0×10^0	100
Untreated seed	–	–	–	–	0	0

– not applicable, *NN, DK* Nor-Natur, Denmark, *ND, DK* Natur Drogeriet, Denmark

[a] Bacterial leaf spot reduction index (BLS-RI) was calculated as $(C - T)/C \times 100$ %, where *C* is the incidence of BLS in tomato seedlings treated with sterile distilled water (negative control) and *T* is the incidence of BLS of tomato seedlings treated with plant extract

[b] Processed coffee (Africafe®) from Afri Tea and Coffee Blenders (1993) Ltd., Dar es Salaam, Tanzania

and cooled for 5 min. The suspensions were filtered through sterile cheesecloth, and the extracts were autoclaved at 115 °C for 15 min and kept at 4 °C until used. Twenty tomato seeds pre-inoculated with *X. perforans*were treated with 1 mL of the 10 % plant extract in an Eppendorf tube and placed on an agitation table at 100 rpm overnight at 25 °C. Untreated, seeds treated with copper sulphate (200 µg/mL $CuSO_4 \cdot 5H_2O$) or sterile distilled were included as controls.

After overnight incubation, the treated seeds were blot-dried and allowed to air-dry for 1 h in the laminar air flow cabinet.

Evaluation of antibacterial activity of plant extracts in vitro

Washing samples (100 µL) from treated seeds were collected using sterile pipettes and serially diluted to 10^{-2} with sterile distilled water in Eppendorf tubes. An aliquot of 100 µL from each dilution was spread onto 0.6 cm thick NA medium in Petri dishes (diameter 8.5 cm × depth 1.3 cm) using a sterile glass rod. The plates were incubated at 28 °C. A pure culture of *X. perforans* NCPPB 4321 was included as a control. Yellow colonies with morphology and color similar to those of *X. perforans* were counted after 96 h. The identity of suspected colonies from each plate was confirmed by pathogenicity tests. The abaxial surfaces of four 14-day-old plants of tomato cultivar Tanya were sprayed to runoff with an inoculum suspension of 10^8 CFU/mL (OD_{600} = 0.01) prepared from 24-h-old bacterial cultures grown on NA at 28 °C. The inoculated seedlings were covered with polyethylene bags and kept in the growth chamber for 14 days. Seedlings sprayed with sterile saline water (containing 0.85 % of NaCl) were used as a negative control, and seedlings sprayed with suspensions prepared from the *X. perforans* culture served as positive controls. The plants were examined for symptoms 14 days after inoculation and scored as negative when no obvious symptoms were observed. Leaves with water-soaked lesions that developed into dark brown spots were scored as positive for BLS disease.

Evaluation of antibacterial activity of plant extracts in planta

To evaluate the effect of the plant extracts on the control of *X. perforans* in seedling assays, 16 tomato seeds treated with plant extracts as previously described were sown in pots (8 cm diameter) containing a 1:3 ratio of sterile sand and peat soil (Pindstrup substrate No. 2, Pindstrup Mosebrug A/S, Ryomgaard, Denmark) and kept in growth chamber at 28 °C under high relative humidity (>85 %). Twenty-one days after sowing, BLS incidence was assessed by calculating the percentage of seedlings with leaf spot symptoms in the total number of emerged seedlings. The efficacy of plant extract treatments in the control of BLS in tomato seedlings was calculated as the BLS reduction index (BLS-RI) = $(C - T)/C \times 100$ %, where C is the incidence of BLS in tomato seedlings raised from seeds treated with sterile distilled water (negative control) and T is the incidence of BLS in tomato seedlings from infected seeds treated with a given plant extract. In addition to the BLS incidence and reduction index, the height, mass and width of seedlings were also evaluated for the best-performing plant extracts. The height of the seedlings was determined by measuring the aerial part of the seedlings from the soil surface to the node of the terminal developing leaf. Fresh mass of the aerial plant part was determined using tomato seedlings cut at the base of the stem by a pair of scissors, and the seedling mass was weighed. To determine the width of the seedlings, we placed a digital caliper at a right angle to the seedlings and recorded the reading at the widest point of the stem. The measurements were repeated in three independent experiments. Plant extracts with the best ability to reduce BLS in tomato during the initial screening tests were determined by comparing the data obtained with Student–Newman–Kuels (SNK) test using SAS version 9.1 software (SAS Institute, Cary, NC, USA). The choice of plant extracts used for screenhouse evaluations in Tanzania was based on four criteria: (1) effectiveness in reducing BLS, (2) most normal seedlings in germination tests, (3) highest vigor index in seedling assays and (4) no phytotoxicity.

Effect of plant extracts on seed germination and seedling growth

The effect of selected plant extracts on seed germination, seedling vigor and mass was evaluated in *Xanthomonas*-free tomato seeds treated with plant extracts as previously described. Seed germination tests were conducted using 400 tomato seeds per treatment. The standard International Seed Testing Association top of paper method (ISTA 2005) was used. The seeds were plated uniformly (50 seeds per replicate) onto three layers of moist blotter paper in a plastic container kept at 27 ± 2 °C and RH >85 % for 14 days. Normal and abnormal seedlings and dead seeds were counted for the germination tests. The same seedlings were used to determine vigor and dry mass. The vigor test involved measurements of root and shoot lengths of seedlings and the percentage seed (normal seedlings) germination. The seedling vigor index (Vi) was calculated as Vi = (mean root length + mean shoot length) × (percentage germination) (Abdul-Baki and Anderson 1973). To determine the dry mass, we wrapped seedlings from each treatment in the germination tests in aluminium foil and dried them in an oven at 103 °C for 24 h. The dried seedlings were then allowed to cool to room temperature and weighed.

Evaluation of plant extracts for production of healthy tomato transplants in the screenhouse

Tomato seeds of cultivars Tanya, Cal-J and Moneymaker, collected from tomato growers in Tanzania and free of BLS-causing xanthomonads, were inoculated with *X. perforans* and treated with plant extracts as previously described. One hundred seeds per cultivar per treatment were sown in polyethylene plastic trays (56.5 × 26.5 × 6 cm) containing a mixture of forest soil, rice husks and farmyard manure (3:1:1). The trays were kept in the screenhouse at 25–33 °C and RH >85 %. Seven days after sowing, 40 seedlings per treatment of each cultivar (10 seedlings per replicate in four replications) were randomly selected and transferred to polyethylene sleeves (6.5 × 9.0 cm) containing the same growth substrate as previously described. The sleeves with the seedlings were placed on the screenhouse benches at the same temperature and RH. BLS incidence and severity in the tomato seedlings were assessed 21 days after sowing. Disease severity was determined based on the Horsfall and Barrett (1945) scale with minor modifications (Shenge 2006), where 1 = no disease, 2 = >0–3 % of leaves with BLS symptoms, 3 = >3–12 % of leaves with BLS symptoms, 4 = >12–25 % of leaves with BLS symptoms, 5 = >25–50 % of leaves with BLS symptoms and 6 = >50 % of leaves with BLS symptoms. In addition to disease incidence and severity, the height and mass of seedlings were also evaluated as already described. The experiment was repeated three times from March to September 2010.

Data analysis

In the in vitro assays, the average of the total number of colony forming units of *X. perforans* on NA was calculated based on three replications. Data for BLS incidence, severity, seed germination, seedling vigor, plant dry and fresh mass and plant height were analyzed using Proc GLM, and mean separation tests were calculated with the Student–Newman–Kuels (SNK) using SAS version 9.1 software (SAS Institute).

RESULTS

Results for preliminary screening of the effectiveness of plant extracts on *X. perforans* indicated that 17 of 84 tested plant extracts (20.2 %) were able to totally reduce the pathogen in in vitro assays (Table 1). The extracts were from *Aloe vera*, *Betula pendula*, processed *Coffea arabica*, *Glycyrrhiza uralensis*, *Juniperus communis*, *Ocimum basilicum*, *Quercus robur*, *Rheum palmatum*, *Rosmarinus officinalis*, *Ruta graveolens*, *Sinapsis alba*, *Yucca schidigera*, *Salvia officinalis*, *Vitis vinifera*, *Punica granatum*, *Olea europea* and *Caryophillus aromaticus*. The effect of these plant extracts on *X. perforans* was similar to that obtained when seeds were treated with copper sulphate (Table 1).

In the in planta assays, 15 of 84 plant extracts (17.8 %) completely inhibited symptoms of BLS in tomato seedlings. Such plant extracts were from *A. vera*, *B. pendula*, *C. arabica*, *G. uralensis*, *J. communis*, *O. basilicum*, *Q. robur*, *R. palmatum*, *R. officinalis*, *R. graveolens*, *S. alba*, *Y. schidigera* and *S. officinallis* (Table 1). Only 13 of 84 extracts (corresponding to 15.5 %) of the assayed plant extracts controlled BLS in both in vitro and in planta assays. These extracts were from *A. vera*, *B. pendula*, *C. arabica*, *G. uralensis*, *J. communis*, *O. basilicum*, *Q. robur*, *R. palmatum*, *R. officinalis*, *R. graveolens*, *S. alba*, *Y. schidigera* and *S. officinalis* and were selected for further experiments.

The results obtained from evaluation of the best-performing plant extracts are shown in Table 2. The 13 selected plant extracts all significantly reduced ($P < 0.001$) the incidence of BLS in tomato seedlings without significantly affecting the growth of tomato seedlings. Seed treatment with plant extracts from *A. vera*, *C. arabica*, *G. uralensis* and *Y. schidigera* totally reduced ($P < 0.001$) the incidence of BLS in tomato (Table 2). The efficacy of these plant extracts to inhibit the growth of *X. perforans* was similar to the effects obtained when seeds were treated with copper sulphate (bactericide) control and untreated (disease free) seed control.

The results also showed that, treatment of tomato seeds with the best-performing plant extracts did not negatively affect the growth of tomato seedlings compared to the treatment with copper sulphate and untreated, disease-free seeds. In contrast, seed treatment with sterile distilled water (negative control) resulted in seedlings with significantly lower ($P < 0.001$) fresh mass (0.31 g) and width (1.45 mm) compared to the other seed treatments (Table 2).

Table 2. Effect of seed treatment with selected plant extracts on incidence of bacterial leaf spot caused by*Xanthomonas perforans* and on growth of tomato seedlings of cultivar Tanya

Treatment	Incidence (%)[a]	Height (cm)	Mass (g)	Width (mm)
Control				
Sterile distilled water	83.30a	13.50b	0.31d	1.45d
Copper sulphate	0.00e	16.13a	0.85abc	1.81abc
Untreated seed	0.00e	15.98a	0.75bc	1.78c
Plant extract				
Aloe vera	0.00e	17.28a	0.96abc	2.05ab
Betula pendula	18.80b	16.90a	0.83abc	2.00abc
Coffea arabica	0.00e	17.13a	0.98ab	2.04ab
Glycyrrhiza uralensis	0.00e	17.23a	0.96abc	2.06ab
Juniperus communis	6.30d	16.11a	0.72c	1.97abc
Ocimum basilicum	18.80b	16.84a	0.78abc	1.93abc
Quercus robur	18.80b	16.69a	0.82abc	1.97abc
Rheum palmatum	18.80b	16.50a	0.82abc	2.04ab
Rosmarinus officinalis	18.80b	16.06a	0.78abc	1.98abc
Ruta graveolens	18.80b	16.16a	0.49ab	1.82abc
Salvia officinalis	12.50c	16.56a	0.82abc	2.04ab
Sinapis alba	12.50c	16.13a	0.74c	1.90abc
Yucca schidigera	0.00e	17.24a	1.00a	2.13a
Mean	12.31	16.47	0.81	1.95
F test	***	***	***	***

*** Significant at $P = 0.01$

[a]BLS incidence = percentage of seedlings with bacterial leaf spot symptoms. Mean followed by same letters in a column are not significantly different based on the SNK test at $P = 0.05$. Each value is a mean of 48 seedlings

The effects of tomato seed treatment with 10 % aqueous plant extracts from *A. vera, C. arabica,G. uralensis* and *Y. schidigera* on seed germination, seedling vigor and dry mass is summarised in Table 3. Treatment of tomato seeds with extracts from *A. vera, C. arabica* and *Y. schidigera*significantly increased ($P < 0.05$) the number of normal seedlings compared to seeds treated with sterile distilled water (negative control). The number of normal seedlings obtained from seeds treated with these extracts was not significantly different from those obtained from copper sulphate (positive control) and untreated tomato seeds (Table 3). The number of normal seedlings obtained with seeds treated with plant extracts of *G. uralensis* was not significantly different ($P < 0.05$) from seed treatments using plant extracts from *A. vera, C. arabica* and *Y. schidigera,* (positive and negative controls, respectively). The number of abnormal seedlings was not significantly different ($P < 0.05$) between different seed treatments (Table 3). When the number of dead tomato seeds was compared between seed treatments, there was no significant difference ($P < 0.05$) between most treatments, except for tomato seeds treated with *C. arabica*(Table 3), which significantly increased ($P < 0.05$) tomato seedling vigor. The treatment of tomato seeds with extracts from *A. vera* and *Y. schidigera* were not significantly different ($P < 0.05$) from seeds treated with *C. arabica*, positive and negative controls (Table 3).

Seed treatment of tomato with plant extracts from *A. vera, C. arabica* and *Y. schidigera*significantly reduced the incidence and severity of BLS ($P < 0.001$) in all three experiments. Such effects were similar to those obtained for seedlings grown from tomato seeds treated with copper sulphate and untreated seeds (disease-free) control (Table 4). In all three experiments, the incidence and severity of BLS disease were significantly higher ($P < 0.001$) in tomato transplants of cultivars Tanya, Cal-J and Moneymaker treated with sterile distilled water (negative control) compared to the other treatments (Table 4).

Table 3. The effect of treatment with selected plant extracts from *Aloe vera*, *Coffea arabica*, *Glycyrrhiza uralensis* and *Yucca schidigera* on tomato seed germination, seedling vigor and dry weight of tomato plants

Treatment	Seed germination[a]			Vigor index (%)[b]	Dry mass (g)
	NS (%)	ABS (%)	DS (%)		
Control					
Sterile distilled water	90.50b	3.00a	6.50a	702.51b	0.16a
Copper sulphate	92.80ab	2.30a	5.00ab	707.60b	0.14a
Untreated seed	92.00ab	3.50a	4.50ab	718.06b	0.16a
Plant extract					
A. vera	94.80a	1.80a	3.50ab	749.39ab	0.14a
C. arabica	95.50a	2.50a	2.00b	784.77a	0.14a
G. uralensis	92.80ab	2.80a	4.50ab	714.33b	0.16a
Y. schidigera	94.30a	2.00a	3.80ab	741.91ab	0.14a
Mean	93.20	2.50	4.30	731.22	0.14
F test	**	ns	**	**	ns

ns not significant

** Significant at $P = 0.05$

[a] Seed germination states: *NS* normal seedlings, *ABS* abnormal seedlings, *DS* dead seed. Each value is the percentage from 400 seeds test

[b] Seedling vigor index (Vi) was calculated as Vi = (mean root length + mean shoot length) × (percentage germination) (Abdul-Baki and Anderson 1973); means followed by the same letters in a column are not significantly different based on the SNK test at $P = 0.05$

DISCUSSION

Tomato seeds were treated with aqueous extracts from 84 different plant materials to assess control of seed-borne infection of BLS of tomato caused by *X. perforans*. In the in vitro assays, 20.2 % of the tested plant extracts totally inhibited growth of *X. perforans* when seed washings from treated seeds were plated on NA. In the in planta experiments, notably 17.8 % of the tested plant extracts reduced BLS incidence by 100 % in tomato seedlings (Table 1). The most effective seed treatments, giving 100 % control in vitro and in planta, were obtained when tomato seeds were treated with plant extracts from *A. vera*, *B. pendula*, *C. arabica*, *G. uralensis*, *J. communis*, *O. basilicum*, *Q. robur*, *R. palmatum*, *R. officinalis*, *R. graveolens*, *S. alba*, *Y. schidigera* and *S. officinalis* (Table 1).

Table 4. The effect of selected plant extracts from *Aloe vera, Coffea arabica* and *Yucca schidigera* applied as seed treatment on incidence and severity of bacterial leaf spot (BLS) caused by *Xanthomonas perforans* on three tomato cultivars under screenhouse conditions in Morogoro, Tanzania

Cultivar	Treatment	Experiment 1		Experiment 2		Experiment 3	
		Incidence (%)[a]	Severity index[b]	Incidence (%)	Severity index	Incidence (%)	Severity index
Tanya	Control						
	Sterile distilled water	45.00a	2.15a	82.50a	2.56a	65.00	2.18a
	Copper sulphate	10.00b	1.10b	0.00b	1.00b	2.50b	1.02b
	Untreated seed	0.00b	1.00b	0.00b	1.00b	0.00b	1.00b
	Extract						
	A. vera	2.50b	1.02b	0.00b	1.00b	0.00b	1.00b
	C. arabica	0.00b	1.00b	0.00b	1.00b	0.00b	1.00b
	Y. schidigera	0.00b	1.00b	0.00b	1.00b	0.00b	1.00b
	Mean	9.58	1.21	13.75	1.26	11.25	1.20
	F test	***	***	***	***	***	***
Cal-J	Control						
	Sterile distilled water	70.00a	2.18a	80.00a	2.55a	60.00a	1.65a
	Copper sulphate	0.00b	1.00b	0.00b	1.00b	0.00b	1.00b
	Untreated seed	0.00b	1.00b	0.00b	1.00b	0.00b	1.00b
	Extract						
	A. vera	2.50b	1.02b	0.00b	1.00b	2.50b	1.02b
	C. arabica	0.00b	1.00b	0.00b	1.00b	0.00b	1.00b
	Y. schidigera	0.00b	1.00b	0.00b	1.00b	0.00b	1.00b
	Mean	12.08	1.20	13.33	1.26	10.42	1.13
	F test	***	***	***	***	***	***
Moneymaker	Control						
	Sterile distilled water	65.00a	2.28a	67.50a	2.40a	72.50a	2.00a
	Copper sulphate	0.00b	1.00b	0.00b	1.00b	0.00b	1.00b
	Untreated seed	0.00b	1.00b	0.00b	1.00b	0.00b	1.00b
	Plant extract						
	A. vera	0.00b	1.00b	0.00b	1.00b	2.50b	1.02b
	C. arabica	2.50b	1.02b	0.00b	1.00b	0.00b	1.00b
	Y. schidigera	0.00b	1.00b	2.50b	1.02b	0.00b	1.00b
	Mean	11.25	1.22	11.67	1.23	12.50	1.17
	F test	***	***	***	***	***	***

*** Significant at $P = 0.01$b

[a] Percentage of seedlings with bacterial leaf spot symptoms

[b] Disease severity index based on Horsfall and Barrett (1945) scale of 1–6 with modifications: 1 = no disease and 6 = >50 % of leaves with BLS symptoms. Means followed by the same letters in a column are not significantly different based on SNK test at $P = 0.05$

From the in planta evaluation of the 13 best performing plant extracts (Table 2), extracts from*A. vera, C. arabica, G. uralensis* and *Y. schidigera* were the most effective and promising for control of BLS of tomato when applied as seed treatment (Table 2). Such results indicate that these plant extracts have bactericidal properties and can be used for tomato seed treatment to control xanthomonads associated with BLS. Many reports are available on the antibacterial properties of these plants. The antibacterial activity of *A. vera* against *Shigella flexneri* and*Streptococcus pyogenes* has been documented (Ferro et al. 2003). Extracts from *A. vera* also affect several Gram-positive and negative bacteria (Cock 2008). The antibacterial activity of extracts from *A. vera* was reported to be due to a direct effect on the bacterial cells caused by the presence of anthraquinones (Boateng 2000) and saponin (Reynolds and Dweck 1999; Urch1999). Other indirect effects have been reported to be associated with the presence of polysaccharides, which stimulate plant defence responses that destroy bacteria (Lawless and Allan 2000; Pugh et al. 2001). Murthy and Manonmani (2009) reported that plant extracts from processed coffee inhibited growth of food-borne pathogens such as *Escherichia coli*, *Yersinia*and *Listeria* species. The antibacterial activity of coffee is associated with substances produced by the roasting process such as millard products, carbohydrate caramelization and thermal composition products (Daglia et al. 1994). In addition, *Glycyrrhiza* species contain α-glycyrrhetinic acid and glycyrrhizin, which inhibit DNA replication and RNA and protein synthesis of microbes (Kim et al. 2002). Other *Glycyrrhiza* species, e.g. *G. glabra*, inhibit growth of some Gram-negative bacteria such as *Salmonella* spp., *Shigella* spp. and *E. coli*(Shirazi et al. 2007). Extracts from *Y. schidigera* also have antibacterial activity, which was attributed to the presence of saponin, a compound found to inhibit microbial growth through hemolytic activity (Hassan et al. 2010). In the present study, observations of cell suspensions of*X. perforans* treated with extracts from *Y. schidigera* at 10 % concentration using confocal microscopy revealed permeabilization of bacterial cells (data not shown).

In addition to reducing BLS in tomato seedlings, plant extracts from *A. vera*, *C. arabica* and *Y. schidigera* significantly improved germination ($P < 0.05$) of tomato seedlings (Table 3). At the same time, the vigor and dry mass of tomato seedlings were not affected by the treatment of seeds with plant extracts from *A. vera*, *C. arabica*, *G. uralensis* and *Y. schidigera*, indicating that these extracts were not phytotoxic to tomato seeds and seedling development. Other seed treatments with natural compounds of plant origin to control plant pathogens have been reported to have no negative effects on plant growth, seed viability or food quality (Opara and Wokocha 2008). Based on data from the present study, extracts from *A. vera*, *C. arabica* and *Y. schidigera* were the most promising plant extracts against BLS and can therefore be used to treat tomato seed. Additionally, they fulfilled the requirements for bioactive chemicals (Hewett and Griffiths 1986) and have the advantages of low mammalian toxicity, minimal health hazards and the least environmental pollution (Amadioha 2003; Singh 1994).

Under screenhouse conditions in Tanzania, using three tomato cultivars, transplants obtained from seed treated with extracts from *A. vera*, *C. arabica* and *Y. schidigera* had significantly lower BLS incidence and severity than tomato seedlings from seed treated with sterile distilled water (Table 4). The ability of these plant extracts to control BLS in the three tomato cultivars grown in Tanzania without negatively affecting seedling growth indicated the potential of using these plant extracts as seed treatment against BLS pathogens.

We also demonstrated that aqueous extracts of 13 plant species (15.5 % of the tested extracts) had antimicrobial properties and inhibited the growth of *X. perforans* in in vitro and in *planta*assays. Plant extracts from *A. vera*, *C. arabica* and *Y. schidigera* were of particular interest as a control strategy to BLS of tomato because they consistently inhibited *X. perforans* in different experiments without negative effects on tomato seeds and seedlings. The ability of these plant extracts to control BLS xanthomonads in tomato also provides an alternative control approach against copper-resistant strains reported to be present in Tanzania (Shenge et al. 2007). These plants extracts must also be tested against BLS under farmers' conditions before they can be recommended for production of BLS free tomato transplants.

More research is also needed to identify bioactive fractions from these plant extracts as well as their mechanisms of action.

ACKNOWLEDGMENTS

The authors wish to thank the Danish International Development Assistance (DANIDA) through ENRECA LIFE project No. 731 for providing funds for this work.

REFERENCES

1. Abdul-Baki AA, Anderson JD (1973) Vigor determination in soybean seed by multiple criteria. Crop Sci 13:630–633

2. Amadioha AC (2003) Evaluation of some plant leaf extracts against *Colletotrichum lindemuthianum* in cowpea. Acta Phytopathol Enthomol Hung 38:259–265

3. Anonymous (2006) List of new names and new combinations previously effectively, but not validly, published. Int J Syst Evol Microbiol 56:925–927

4. Black R, Seal S, Abubakar Z, Nono-Womdim R, Swai I (2001) Bacterial spot (*Xanthomonas campestris* pv. *vesicatoria*) of tomato and sweet pepper in Tanzania. Plant Pathol 50:810 (abstract)

5. Boateng JS (2000) Analysis of commercial samples of aloe. Ph.D. thesis, University of Strathclyde, Glasgow

6. Carrillo-Fasio JA, Garcia-Estrada RS, Allende-Molar R, Marquez-Zequera I, Millan- Ocampo S, Gaxiola-Espinoza G et al (2001) Sensitivity of *Xanthomonas campestris* pv. *vesicatoria*(Doidge) dye strains to copper. Rev Mex Fitopatol 19:72–77

7. Cock IE (2008) Antimicrobial activity of *Aloe barbadensis* Miller leaf gel components. Internet J Microbiol 4(2). http://www98.griffith.edu.au/dspace/bitstream/10072/21381/1/50582_1.pdf. Cited 12 April 2011

8. Daglia M, Cuzzoni MT, Dacarro C (1994) Antibacterial activity of coffee. J Agric Food Chem 42:2270–2272

9. Ferro VA, Bradbury F, Cameron P, Shakir E, Rahman SR, Stimson WH (2003) In vitro susceptibilities of *Shigella flexneri* and *Streptococcus pyogenes* to inner gel of *Aloe barbadensis*Miller. Antimicrob Agents Chemother 147:1137–1139

10. Gitaitis R, McCarter S, Jones J (1992) Disease control in tomato transplants produced in Georgia and Florida. Plant Dis 76:651–656

11. Gore JP, O'Garro LW (1999) *Xanthomonas campestris* pv. *vesicatoria* from bell pepper and tomato in Barbados undergoes changes in race structure, virulence and sensitivity to chemical control agents. J Phytopathol 47:397–402

12. Hassan SM, Byrd JA, Cartwright AL, Bailey CA (2010) Hemolytic and antimicrobial activities differ among saponin-rich extracts from guar, quillaja, yucca, and soybean. Appl Biochem Biotechnol 162:1008–1017

13. Hewett PD, Griffiths DC (1986) Biology of seed treatment. In: Jeffs KA (ed) Seed treatments. BCPC Publications, Thornton Health, Surrey, pp 7–12

14. Horsfall JG, Barrett RW (1945) An improved grading system for measuring plant disease. Phytopathology 35:655 (abstract)

15. International Seed Federation (ISF) (2007) Method for the detection of *Xanthomonas campestris* pv. *vesicatoria* on tomato seed. ISF Secretariat, Chemin du Reposoir, Nyon

16. International Seed Testing Association (ISTA) (2005) International rules for seed testing. Bassersdorf

17. Jones JB, Bouzar H, Stall RE, Almira EC, Roberts PD, Bowen BW, Sudberry J, Strickler PM, Chun J (2000) Systematic analysis of xanthomonads (*Xanthomonas* spp.) associated with pepper and tomato lesions. Int J Syst Evol Microbiol 50:1211–1219

18. Jones JB, Lacy GH, Bouzar H, Stall RE, Schaad NW (2004) Reclassification of the xanthomonads associated with bacterial spot disease of tomato and pepper. Syst Appl Microbiol 27:755–762

19. Kaaya NKF, Mortensen CN, Mabagala RB, Massomo SMS (2003) A guide on seed-borne bacterial diseases of tomato in Tanzania. Technical Bulletin, Danish Government Institute of Seed Pathology for Developing Countries (DGISP), Copenhagen, Denmark

20. Kim HK, Park Y, Kim HN, Choi BH, Jeong HG, Lee DG, Hahm KS (2002) Antimicrobial mechanism of β-glycyrrhetinic acid isolated from licorice, *Glycyrrhiza glabra*. Biotechnol Lett 24:1899–1902

21. Kumar J, Parmar BS (1996) Physicochemical and chemical variation in neem oils and some bioactivity leads against *Spodoptera litura* F. J Agric Food Chem 44:2137–2143

22. Lawless J, Allan J (2000) The chemical composition of *Aloe vera*. In: *Aloe vera* natural wonder cure. Thorsons Publishing, London, pp 161–171

23. Lee SD, Cho YS (1996) Copper resistance and race distribution of *Xanthomonas campestris* pv.*vesicatoria* on pepper in Korea. Korean J Plant Pathol 12:150–155

24. Martin HL, Hamilton VA, Kopittke RA (2004) Copper tolerance in Australian populations of*Xanthomonas campestris* pv. *vesicatoria* contributes to poor field control of bacterial spot of pepper. Plant Dis 88:921–924

25. Murthy PS, Manonmani HK (2009) Physico-chemical, antioxidant and antimicrobial properties of Indian monsooned coffee. Eur Food Res Technol 229:645–650

26. Myung IS, Jeong IH, Moon SY, Lee SW, Shim HS (2010) A new disease, arboricola leaf spot of bell pepper, caused by *Xanthomonas arboricola*. Plant Dis 94:271(abstract)

27. Opara EU, Wokocha RC (2008) Efficacy of some plant extracts on the in vitro and in vivo control of *Xanthomonas campestris* pv. *vesicatoria*. Agric J 3:163–170

28. Prakash A, Rao J (1997) Botanical pesticides in agriculture. Lewis Publishers, London

29. Pugh N, Ross SA, ElSohly MA, Pasco DS (2001) Characterization of aloeride, a new high-molecular-weight polysaccharide from *Aloe vera* with potent immunostimulatory activity. J Agric Food Chem 49:1030–1034

30. Punina NV, Ignatov AN, Pekhtereva ESH, Kornev KP, Matveeva EV, Polityko VA, Budenkov NI, Schaad NW (2009) Occurrence of *Xanthomonas campestris* pv. *raphani* on tomato plants in the Russian Federation. Acta Hortic 808:287–290

31. Reynolds T, Dweck AC (1999) *Aloe vera* leaf gel: a review update. J Ethnopharmacol 68:3–37

32. Scheck HJ, Pscheidt JW, Moore LW (1996) Copper and streptomycin resistance in strains of*Pseudomonas syringae* from Pacific Northwest nurseries. Plant Dis 80:1034–1039

33. Shenge KC (2006) Bacterial speck and spot diseases of tomato in Tanzania: pathogen characterization, epidemiology and management options. PhD thesis, Sokoine University of Agriculture, Morogoro

34. Shenge KC, Mabagala RB, Mortensen CN (2007) Identification and characterization of strains of *Xanthomonas campestris* pv. *vesicatoria* from Tanzania by biolog system and sensitivity to antibiotics. Afric J Biotech 6:015–022

35. Shirazi MH, Ranjbar R, Eshraghi S, Sadeghi G, Jonaidi N, Bazzaz N, Izadi M, Sadeghifard N (2007) An evaluation of antibacterial activity of *Glycyrrhiza glabra* extract on the growth of*Salmonella*, *Shigella* and ETEC *E. coli*. J Biol Sci 7:827–829

36. Singh DC (1994) Scope of medicinal and aromatic plants in pest management. In: Narwal SS, Tauro P (eds) International symposium, allelopathy in sustainable agriculture, forestry and environment, New Delhi, p 68

37. Stall RE, Beaulieu C, Egel D, Hodge NC, Leite RP, Minsavage GV, Bouzar H, Jones JB, Alvarez AM, Benedict AA (1994) Two genetically diverse groups of strains are included in *Xanthomonas campestris* pv. *vesicatoria*. Int J Syt Bacteriol 44:47–53

38. Tamir-Ariel D, Navon N, Burdman S (2007) Identification of genes in *Xanthomonas campestris*pv. *vesicatoria* induced during its interaction with tomato. J Bacteriol 189:6359–6371

39. Urch D (1999) Aloe vera nature's gift. Blackdown Publications, Bristol

Chapter 9

Effect of Antagonists and Plant Extracts in the Control of Protea Wilt (F. oxysporum)

Edgar Martínez Granja*, **Sergio Reyes Benitez, Danny Sanjuanello**

Facultad de Ingeniería Agronómica, Universidad de Ciencias Aplicadas y Ambientales, Bogotá, Colombia

ABSTRACT

In laboratory experiments were evaluated TRICHOD®, FOLIGUARD®, Trichoderma sp, and watery extracts of pine, marigold and eucalyptus, respectively, against F. oxysporum. There were no significant differences between TRICHOD® and FOLIGUARD®. The extract of marigold of 120 g/L was the best result in reducing the area of the colony of F. oxysporum. In field pots experiment were evaluated the extract of marigold (120 g/L), potassium phosphite (400 g/L), TRICHOD and be- nomyl (1.5 g/L) to control F. oxysporum. Treatments were applied to the soil, 15 days after inocu- lation with F. oxysporum. The best obtained heights were of 32 cm with extract of marigold 120 g/L, followed by potassium phosphite with 31 cm, and the control reached 26 cm of height. The treatment with an extract of marigold 120 g/L, obtained the best dry weight average of 6.5 g, sta- tistically different from the control. The extract of marigold presented the best efficiency against the disease, because it reduced the length of the vascular symptom in 88.5%; followed by TRICHOD®, benomyl and potassium phosphite that showed reductions of 86, 85 and 84 percent, respectively.

Keywords: Marigold Extract, Plant Disease Control, Trichoderma

1. INTRODUCTION

The diversification in planting ornamental species has been a strategy used by the producers, as an alternative to extending markets, to improve competitiveness and to increase profitability. A step in this respect has been the production and commercialization of new flowers and foliages, as the proteas of the genus Leucadendron.

The cultivation of proteas is relatively new in the market of ornamental plants for the production of flowers, dried flowers and flower pots [1] . There is great diversity of proteas and the genus Leucadendron has stand out in international trade, mainly in the European community, as cut and processed as dried flower blossom. The main consumers of proteas are Japan, Europe and the United States. This market has an estimated value of 455 million euros. The total number of stems exported around the world is 100 million units and about half of total world exports are of the genus Leucadendron [2] .

Proteas production has been affected by the incidence of diseases caused by three pathogens of great impor- tance that inhabit the soil: Phytophthora cinnamomi, Rhizoctonia spp. and Fusarium spp. [3] - [5] . In the genus Fusarium three species are agents of Leucadendron wilt: F. oxysporum, F. solani and F. moniliforme. F. oxys- porum was first detected in the northwestern region of South Africa and Zimbabwe [6] . It has subsequently been reported in different parts of the world where proteas are cultivated and it was reported the incidence of F. solani and F. equiseti attacking proteas of the genus Leucandendron especially in the Madeira Island in Portugal [3] . It has also been found in the Canary Islands [7] and in Australia [8] . In Australia, was found F. oxysporum in roots, neck and vascular tissue of plants with wilt symptoms [9] . Also, has been detected in seedlings and com- mercial plantations, causing losses greater than 60% [8] - [10] .

In Colombia, F. oxysporum is the causal agent of Leucadendron wilt [1] . Symptoms have been observed in young and adult plants of the protea cultivars Petra and Gold Strike. Necrosis was observed at the beginning in young leaves and side shoots that die after infection, generating a reduction in the number of flowers produced per plant [6] - [8] . Vascular tissues acquire a dark brown color which extends from the root to the stem of the plant and affects both the xylem and phloem. There is a turgor loss in the apical part of the shoot, yellowing oc- curs from the lower leaves towards the top of the branch. Subsequently, the branches weaken and necrosis ap- pears in leaves and the stem [1] .

In general, the diseases caused by F. oxysporum are difficult to control. Resistant varieties, planting certified seeds, and healthy seedlings are the best practices to control wilt caused by Fusarium species. Also antagonist biocontrol agents, such as Trichoderma, offer possibilities to control, primarily because it inhibits the growth and development of the disease [11] . The genus Trichoderma has fast reproduction; it is able to survive and to modify the rhizosphere and it is effective in promoting plant growth [12] . The incidence of proteas wilt was re- duced to 20% by using a native strain of T. harzianum in concentrations of 10^8 and 10^9 conidias per mL [7] . Tricoderma has been commercially produced to be used in controlling several plant pathogens such as: Pythium spp, Rhizoctonia spp, Sclerotium spp, and Fusarium spp [13] .

A variety of Trichoderma species has been evaluated to control F. oxysporum in different crops. To control this pathogen in papaya plants (Carica papaya L.) a treatment of Trichoderma at a concentration of 10^6 conidia wasapplied to the soil artificially infested with F. oxysporum. Its antagonist effect has also been evaluated on tomato plants to wilt caused by F. oxysporum [14] .Tomato seeds coated with T. harzianum were planted in nat- ural soil artificially infested with F. oxysporum [15] . The seed treatment caused no reduction in the disease, whereas the soil treatment was able to reduce 92% of the infection. In Cuba T. harzianum applied 20 g/plant in soils with occurrence of Panama disease, provided superior control to 95% in the varieties tested [16] .

In searching for new alternatives to organic plant disease management, the use of plant metabolites has been considered feasible [17] . Botanical pesticides are chemical compounds that occur naturally in plants. They have been used to reduce the growth of certain fungi under laboratory conditions. Extracts of eucalyptus and other plants in different concentrations reduced the growth of F. solani. The marigold (Calendula officinalis) is culti- vated in different countries, and some studies have been developed using this plant extractto control some species of nematodes that attack citrus, but its greatest use has been in human dermatology and food industry, due to its antimicrobial properties [18] [21] . Essential oils are extracted from roots, seeds, leaves and flowers of marigold, which are rich in sesquiterpenes, phenols and saponins that are responsible for the antifungal activity [22] . Sesquiterpenes activity was evaluated in vitro for control C. acutatum, C. fragariae, C. gloesporoides, F. oxysporum, Botrytis cinerea, and Phomopsis sp [23] .

In Colombia, the growers apply some chemical fungicides, such as benomyl to wilt control. This product re- duced by 50% the incidence of wilt caused by F. oxysporum on susceptible tomato plants [24] . Similar results were obtained in tomato [25] and susceptible native bean varieties planted in soil naturally infested with Fusa- rium spp and Rhizoctonia solani [26] . Also it reduced the Panama disease (F. oxysporum. f. sp. cubense) in ba- nana by 70% with a drench application one week after transplantation was made [27] . According to Obreque [7] benomyl has been widely used in the preventive control of wilt of proteas in South Africa.

The antimicrobial properties of the phosphites were found in studies aimed at controlling diseases caused by oomycetes [28] [29] . Also, Mogollón and Castaño [30] , showed the effect of potassium phosphite as resistance inductor, when it was applied to the soil to control sigatoka disease caused by M. fijiensis and M. musicola in plantains. And Walters et al., [31] demonstrated the effectiveness of potassium phosphiteto control different diseases, including those caused by Phytophthora infestans, Fusarium oxysporum and Rhizoctonia solani. It was found that the foliar application of potassium phosphite reduced by 40% the severity of wheat wilt caused by F. culmorum [32] .

The main objective of this research was to determine the effect of Trichoderma spp, potassium phosphite, benomyl and plant extracts in the control of F. oxysporum in Leucadendron proteas, in laboratory and field pot conditions.

2. MATERIALS AND METHODS

2.1. Laboratory Experiments

F. oxysporum was isolated from diseased Leucadendron proteas collected in established crops, in a farm located in Subachoque, Department of Cundinamarca, Colombia, at 2600 altitude, and temperature between 15°C and 19°C. Wilty Petra variety plants were chosen with external and internal symptoms. The fungus was isolated in acidified Potato Dextrose Agar (PDA + A) with 2500 μL of 50% lactic acid per 1000 mL of medium. Stem samples with visible wilt symptoms were previously washed with tap water to remove soil residues. Of each stem, square 5 mm pieces were cut, and immersed 1 min in 70% ethanol, two minutes in sodium hypochlorite 1% and washed in sterile distilled water for 30 s [33] . In a laminar flow cabinet, four pieces were planted on each Petri dish containing PDA + A using a sterilized forceps. The isolated samples were incubated at 20°C for eight days. On slides, stained with lactophenol cotton blue, were examined at light optical microscope (Nikon eclipse E 100) at 40× and 100× to check the morphological characteristics of the pathogen, by comparison with diagrammatic keys [34] [35] .

Data from each of the two laboratory experiments, were processed with the SAS statistical program, through which the fulfillment of assumptions prior to the execution of the analysis of variance was tested. In cases where the F-test analysis of variance was significant, the honest Tukey multiple comparison (Tukey HSD) was used to determine the differences within treatments.

Experiment 1. The TRICHOD® (Orius Biotecnologia) and FOLIGUARD® (Live Systems Technology) commercial biological products with the active ingredient Trichoderma harzianum and an isolate of Trichoderma sp uncharacterized, were evaluated against F. oxysporum using spore suspensions at a concentration of 1×10^8 spores/mL. A completely randomized design was followed, with 15 replications, where each treatment cor- responded to a biological product. Two discs of 8 mm diameter, one of the pathogens and one of the antagonists, were transferred to a Petri dish with PDA + A. These discs were located one in front of the other at a distance of 3 cm, and were incubated at 20°C. The pathogen area growth in each treatment, and the control without treat- ment was measured three times at week for a period of three weeks.

Experiment 2. In order to set the action of plant extracts of pine, eucalyptus and marigold on the develop- ment of colonies of F. oxysporum, the next procedure was followed: Samples of these plant species were col- lected in the campus of UDCA, washed with water, immersed in sodium hypochlorite 1% for two minutes, and finally rinsed with sterile distilled water. Three subsamples of 60 g, 120 g and 180 g, for each one of pine, euca- lyptus and marigold samples, respectively, were evaluated. Each subsample was boiled in a liter of sterile dis- tilled water and subsequently filtered through filter paper, adjusted to 1000 ml of water to prepare a culture me- dium PDA + A + Extract (PDA + A + E).

From the pure culture of F. oxysporum obtained in experiment 1, 15 discs of 8 mm in diameter were extracted and planted in individual petri dishes containing the PDA + A + E culture medium, and incubated at

20°C. The colony area was measured three times a week for a period of three weeks. It was followed a completely rando- mized design with 15 replications for each extract and concentration.

2.2. Field Pots Experiment

Leucadendron proteas plants, cultivar "Safari Sunset", one month old at the time of transplantation, with no in- ternal and external disease symptoms were planted in 5 kg pots, filled with unsterilized soil obtained in the ex- perimental field of UDCA. Once planted they were located according to the experimental design in rows of 10 pot plants per each of the five treatments, spaced 0.30 m apart.

Then, a randomized complete block design was followed with five treatments and five replications. The treat- ments were T. harzianum, marigold plant extract 120 g/L, potassium phosphite, benomyl 50 WP, and the absolute control. The treatments had randomization within the blocks and in the experimental area. Each experimental unit consisted of 10 plants for a total of 250 plants. For data analysis, the Minitab 16 statistical software was used.

To isolate F. oxyporum it was followed the procedure previously described. Ten days after planting, the seedlings were inoculated with a suspension of 10^6 macroconidia/mL of F. oxysporum [Salazar et al., 2010]. One mL of the suspension was injected into the root collar, using a plastic syringe. Fifteen days after the inocu- lation the following treatments were applied: 1) 5 mL per pot around the plant of a suspension of TRICHOD® at a concentration of 2×10^{10} conidia/mL of T. harzianum; 2) This treatment consisted of the application of 10 mL of marigold extract, around the neck of the plant. The marigold extract was prepared with 120 g of marigold leaves per liter of water, because the best results were obtained in the laboratory study; 3) For the treatment with the fungicide the product benomyl 50 WP was used in doses of 1.5 g per liter of water. Two mL of the solution were used to drench the soil around the stem in each plant; 4) To evaluate the potassium phosphitethe commer- cial product Agrifos 400 SL, was used. The drench had a concentration of 7.5 mL of the commercial product per liter of water. Each plant was drenched with 5 mL of this solution; 5) The control did not receive chemical nor biological treatments.

For six months, every 15 days data were taken in five randomly selected plants in each treatment and replica- tions. a) The height of the plant was measured between the soil surface and the apex of the last leaf of the plant with a ruler; b) The number of side shoots was quantified; c) At the end of six months, no external symptoms of the disease manifested and for this reason destructive sampling was performed in order to observe wilt symp- toms in the vascular bundles of the plant. Dry matter accumulation was determined in the roots of the plants se- lected at random, in the destructive sampling. Fresh weight was taken and then placed in an oven for 48 hours at 60°C, to obtain dry weight. Five plants per treatment were evaluated, totaling 25 plants. For data analysis the Minitab 16 statistical software was used and the data were subjected to analysis of variance, Tukey test and the examination of contrasts.

3. RESULTS AND DISCUSSION

3.1. Experiments at Plant Pathology Laboratory

Significant differences were found ($p < 0.01$) among three treatments: TRICHOD®, FOLIGUARD®and Tri- choderma sp, and the control. In general, they reduced the growth of the F. oxysporum colony. Duncan test in- dicates that TRICHOD® reduced significantly the colony area of F. oxysporum, followed by FOLIGUARD® and Trichoderma sp (Figure 1).

The results indicate that Trichoderma inhibits colony growth of F. oxysporum under laboratory conditions. This antagonistic effect has already been described by other researchers to explain the potential of

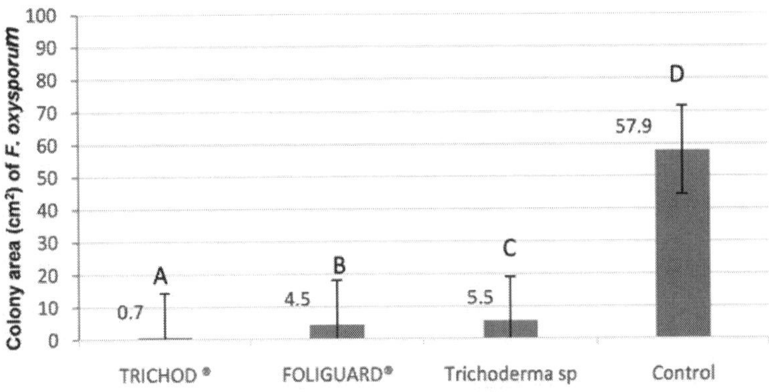

Figure 1. Effect of TRICHOD®, FOLIGUARD® and Trichoderma sp on the colony area of F. oxysporum. Means of 15 replications. Means with the same letter are not significantly different by the Duncan test.

the genus Trichoderma for the management of this important plant pathogen. Cifuentes [36] found that T. harzianum was highly effective in controlling Fusarium solani in tomato. Also, when the chemical control and the use of anta- gonists against F. graminearum were compared, it was determined that copper oxychloride inhibited growth of the pathogen, whereas Trichoderma reduced 64% the colony growth and was effective in reducing the severity of the disease in wheat crops [37] . Meanwhile, it was established [16] that T. harzianum obtained 95% control in plantations of banana affected by Panama disease.

According to Stefanova et al., [38] and Selosse et al., [39] , the activity and biocontrol efficacy of Trichoderma is based on the property to act directly as hyperparasite, and indirectly by means of the action of antifungal meta- bolites and hydrolytic enzymes that cause cellular changes in the pathogen, and may cause degradation of the cell wall of F. oxysporum. These characteristics of Trichoderma will increase their competitiveness and effec- tiveness in controlling other pathogens that inhabit the soil.

The tested plant extracts showed significant differences ($p < 0.01$) among treatments (Figure 2) and reduced the colony area of F. oxysporum and best results were found with marigold extract. The three concentrations evaluated were significantly different from other treatments and the control, but no statistical differences were found between the three concentrations of marigold. Treatment corresponding to 120 g, had the lowest growth area of F. oxysporum. No differences were found between pine and eucalyptus extracts and the control.

There is a possibility of using plant extracts to control plant pathogens. In coffee plants promising results have been obtained in the management of diseases because of their mechanism of direct action on pathogens, and their potential for inducing systemic resistance [40] - [43] . The antimicrobial action of marigold extract is linked to the production of secondary metabolites content in vegetables such as flavonoids, salicylic acid, sesquiter- penes and phenols [23] . For example, phenols, are toxic against Fusarium species [41] . Eucalyptus extract has been found to reduce the growth of mycelium of Colletotrichum, but no effects on Fusarium are known [42] .

The results explore the possibility of new organic technologies that allow a sustainable management of plant health problems, which counteract the harmful effects of the use of synthetic chemical fungicides. New tech- nologies are a common benefit to produce proteas and other crops where Fusarium attacks generate large losses. Moreover, the results are a significant contribution to knowledge and use of biodiversity, because working with native and introduced plants, and domesticated microorganisms in Colombia generate knowledge to expand their potential uses.

3.2. Field Pots Experiment

Plant height. The data were transformed, and significant differences ($p < 0.001$) among treatments and blocks were found. The heights were 31 cm and 32 cm between plants treated with potassium phosphite and plants treated with marigold extract, respectively. These results were statistically different to the control (Figure 3). Theapplication of benomyl, could have an initial effect on the incidence of the disease, but was not persistent enough to allow normal plant development. The control presented a marked delay in growth with 26 cm and development compared to treated plants.

Number of side shoots. There were no significant differences ($p \leq 0.05$) among the treatments: TRICHOD, potassium phosphite, marigold extract and benomyl. However, there were differences between these treatments and the control. The greatest number of side shoots was 2.5 occurred with the TRICHOD® treatment (Figure 4). Avis et al., (2008) highlights the action of Trichoderma not only as a biocontrol agent, but also as a growth promoter. Different strains of T. harzianum and T.

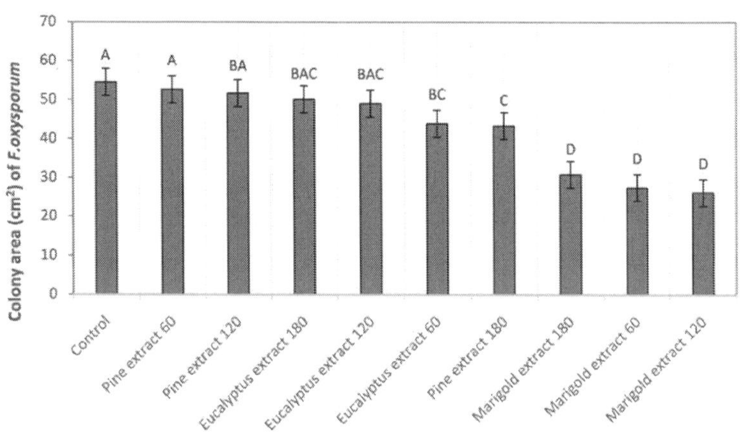

Figure 2. Effect of pine, eucalyptus and marigold extracts on the colony area of F. oxysporum. Means of 15 replications. Means with the same letter are not significantly different by the Duncan test.

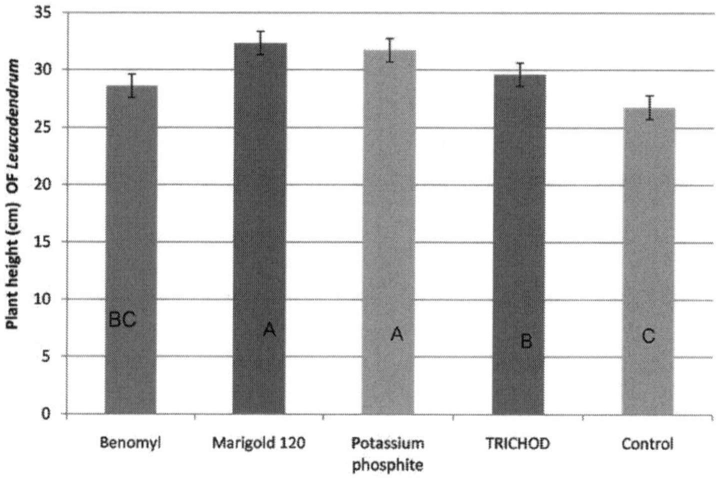

Figure 3. Effect of benomyl, marigold 120, potassium phosphite and TRICHOD on plant height of Leucadendrum. Means of 5 replications. Means with the same letter are not significantly different by the Tukey test.

viride have shown increases in fresh and dry weights of later- al stems; also the increase in volume and dry weight of roots have been found in vegetables and other crops such as tobacco. In addition, two mechanisms of action have been identified with respect to the benefits of Tricho- derma on plant growth. The first is an increase in nutrition through the solubilization and absorption of micro and macro nutrients, and the second is production-related growth promoters in plants, such as indol acetic acid [43] . Side effects of the antagonist found in this research are important because the productivity of cut flowers is measured by the number of lateral stems with flower. In plantations, it is possible to reach better performance with this treatment, even with some incidence of the disease. Marigold treatments, potassium phosphite and be- nomyl, were, 2.5, 2.3 and 2.4 side shoots per plant respectively, exhibiting difference versus the lowest average of 2.2 lateral shoots per plant. In addition, there is a confirmation of the effect of F. oxysporum on the develop- ment and growth of the plant when it does not receive any treatment.

3.3. Root Dry Weight

Treatment with marigold extract 120 g/L, had the best average of 6.5 g and it was statistically different from the control (Figure 5), but there was no difference between the marigold, potassium phosphite, benomyl and TRICHOD® treatments. When the dry matter accumulation in the root and the average plant height are com- pared, similar behaviors are observed. A larger and well-structured root may provide greater anchorage and nu- trient uptake to the plant, allowing better growth and development (Figure 5). F. oxysporum causes damage to the root, affects the amount of functional roots and decreases the amount of water absorbed by these. Some vas- cular pathogens are related to the inhibition in the production of root hairs, which reduces the absorption [11] .

3.4. Disease Incidence

The disease did not show any external symptoms after six months of performing the inoculation. However, when the destructive sampling was done, it was found 100% incidence in all treatments. In the destructive sam pling, it was evident the manifestation of symptoms in the vascular bundles, characterized by brown color around the xylem, extending from the base of the stem to the lateral side stems.

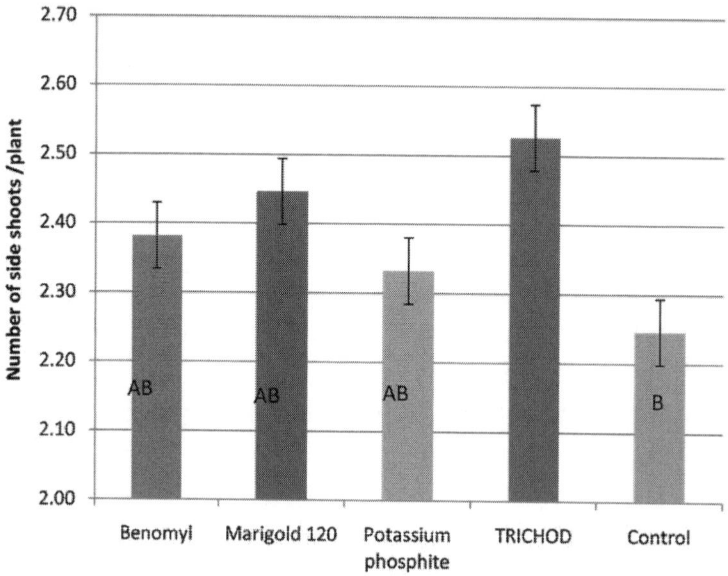

Figure 4. Effect of benomyl, marigold 120, potassium phosphite and TRICHOD on the num- ber of side shoots of Leucadendrum plants. Means of 5 replications. Means with the same letter are not significantly different by the Tukey test.

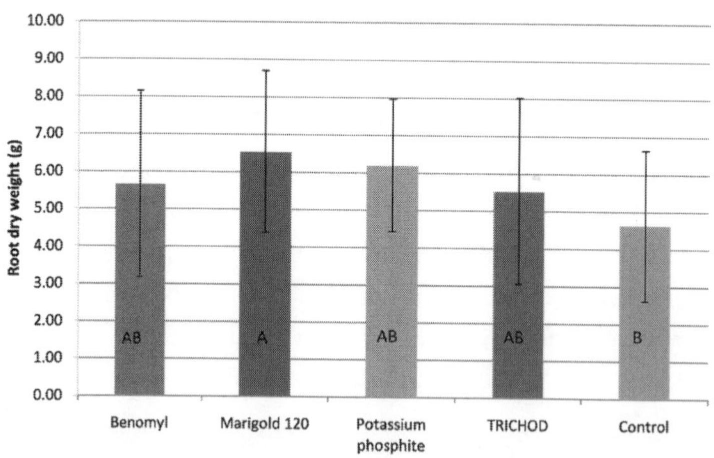

Figure 5. Effect of benomyl, marigold 120, potassium phosphite and TRICHOD on the root dry weight of Leucadendrum plants. Means of 5 replications. Means with the same letter are not significantly different by the Tukey test.

There were significant differences ($p \leq 0.001$) between treatments. The marigold extract presented the best efficacy against the disease, because reduced damage in 88.5%, followed by TRICHOD, benomyl and potassium phosphite, which showed reductions in 86%, 85% and 84%, respectively. Although the treatment of potassium phosphite was not superior to marigold extract, it did show good plant height average (Figure 3), and it was able to reduce the development of disease. This result could be related to the induction of resistance mechanisms that allowed the plant to generate tolerance to the disease, and continue their growth and development. These results agree with the work of Monsalve et al., [44] who evaluated the effects of induced phytoalexins from phosphites in Oomycetes, Rhizoctonia solani and Botrytis cinerea.

Overall, marigold extract produced the best results in this investigation. Marigold has a number of compounds, such as sesquiterpenes, sesquiterpenoles, saponins and flavonoids that give certain anti-fungal properties. Mar- tinez [22] mentions that flavonoids are aromatic compounds possessing anti-fungal activity against pathogens such as Penicillium sp and Rhizopus sp, amongst others. Phenolic compounds pervade pathogen cell membranes, resulting in leakage of cytoplasmic contents. According to Wedge et al. [23] , sesquiterpenes are a family of more than 5000 compounds found primarily in members of the Compositae family. Sesquiterpenes have a broad spectrum of biological activity that plays an important role in the defense mechanisms of the plant. The same author found that sesquiterpene lactone had an inhibitory action against Colletotrichum, Phomopsis, Botrytis, but not against F. oxysporum isolated from Leucadendron sp.

Treatment with T. harzianum did not provide outstanding averages in the expression of growth traits, except the number of side shoots, which differed from the others; however, presented a significant reduction in lesion length, compared with the control. Possibly, the application of T. harzianum must be done before the infection, for greater efficiency, because, as demonstrated in the results obtained by

Perez et al. [16] , there was a complete control of F. oxysporum in banana plants treated with T. harzianum.

In contrast test was found that the length of the lesion was significantly higher in benomyl treatment versus marigold extract, indicating that the latter was better than the chemical treatment. Studies developed by Amini and Sidovich [25] indicate that better control of F. oxysporumf. sp. lycopersici was achieved with the application of pre-infection fungicide. This explains that the time of application affects the fungicide control responses of the disease. However, it is different with results obtained by Obreque [7] , who obtained a 100% decrease in the Leucandendron disease when it was applied 30 days after inoculation. This result could be associated to the va- riability of F. oxysporum or the fungicide application method used.

ACKNOWLEDGEMENTS

Edgar Martínez Granja, Sergio Reyes Benitez, Danny Sanjuanello To Universidad de Ciencias Aplicadas y Ambientales UDCA, for supporting this research.

REFERENCES

1. Salazar, I., Martínez, E., Peña, F., Cabezas, M. and Gómez, J. (2010) Identificación del agente causante de la marchitez de proteas (Leucadendronsp.). Rev. U.D.C.A. Act. & Div. Cient., 13, 79-87.

2. Littlejohn, G.M. (2001) The Challenges of Breeding Wild Flower Cultivars for Use in Commercial Floriculture: Afri- can Proteaceae. 20th International Eucarpia Symposium, Section Ornamentals, Strategies for New Ornamentals—Part I 552, 25-38.

3. Moura, M. and Rodríguez, P. (2001) Fungal Diseases on Proteas Identified in Madeira Island. ISHS ActaHorticulturae 545: V International Protea Research Symposium, 265-268.

4. Denman, S. (2001) Evaluation of a Stem Inoculation Technique for Assessing Resistance to Phytophora cinnamomi in Leucadendron y Leucospermun Cultivars. ISHS ActaHorticulturae, 545, 281-285.

5. Brits, G. and Broembsen, S. (1985) Effects of Tillage on Heath Land Soil. Disturbance of a Natural Mountain Fybos System in the Southwestern Cape. South Africa. Soil and Tillage Research, 24, 29-45.

6. Swart, L., Demman, S., Lamprecht, P. and Crous, P. (2001) Fusarium Wilt: A New Disease of Cultivated Protea in South Africa. Australasian Plant Pathology, 258, 156-161.

7. Obreque, X. (2004) Evaluación de aplicaciones preinfección del fungicida benomylo y del biocontrolador Tricho-dermaharzianum en el control de Fusarium sp en Proteaceae. Escuela de Agronomía. Universidad de Talca, Chile. http://dspace.utalca.cl/ [Paper reference:4]

8. Lubbe, C., Lamprecht, S., Van Niekerk, J. and Mostert, L. (2008) Molecular Characterization of Fusariumoxysporum Causing Wilt of Proteaceae. ISHS ActaHorticulturae, 805, 127-134.

9. Dunne, C., Dell, B. and Hardy, G. (2003) "Sudden Death" in Proteas in the Southwest of Western Australia. ISHS Acta-Horticulturae, 602, 39-44.

10. Benic, L.M. (1986) Pathological Problems Associated with Propagation Material in Proteaceae Nurseries in South Africa. ActaHort, 185, 229-236.

11. Agrios, G.N. (2005) Plant Pathology. 5th Edition, Elsevier Academic Press, Amsterdam. [Paper reference:2]

12. Pandya, J.R., Sabalpara, A.N. and Chawda, S.K. (2011) Trichoderma: A Particular Weapon for Biological Control of Phytopathogens. Journal of Agricultural Technology, 7, 1187-1191.

13. Intana, W. (2003) Selection and Development of Trichoderma spp. for High Glucanasa, Antifungal Metabolite Pro- ducing and Plant Growth Promoting Isolates for Biological Control of Cucumber Damping-Off Caused by Pythium spp. Ph.D. Thesis, Kasetsart University, Bangkok.

14. González, J.C., Maruri, J.M. and González, A. (2005) Evaluación de diferentes concentraciones de Trichodermaspp contra Fusarium oxysporum agente causal de la pudrición de plántulas en papaya (Carica papaya L.) en Tuxpan, Veracruz, México. Revista UDO Agrícola, 5, 45-47.

15. Ramezani, H. (2010) Antagonistic Effects of Trichoderma spp. against Fusariumoxyxporumf. sp. lycopersici Causal Agent of Tomato Wilt. Plant Protection Journal, 2, 167-173.

16. Pérez, L., Batlle, A., Chacón, J. and Montenegro, V. (2009) Eficacia de Trichodermaharzianum A34 en el biocontrol de Fusarium oxysporumf. sp. cubense, agente causal de la marchitez por Fusariumo mal de Panamá de los bananos en Cuba. Fitosanidad, 13, 259-264.

17. De Marcano, A., Vargas, D.N. and Pire, A.(2005) Efecto de extractos vegetales y fungicidas sintéticos sobre el cre- cimiento micelial in vitro de Sclerotiumrolfsiiy Thielaviopsisbasicola. Revista Facultad de Agronomía, 22, 315-323.

18. Joseph, B., Dar, A. and Kumar, V. (2008) Bioefficacy of Plantextractsto Control Fusarium solani F. Sp. Melongenae

Incitant of Brinjal Wilt. Global Journal of Biotechnology & Biochemistry, 3, 56-59.

19. Montasser, S.A., El-wahab, A.E., Abd-Elgawad, M.M., Abd-El-Khair, H., Koura, F.H. and Hammam, M.M. (2012) Role of Some Plant Extracts and Organic Manure in Controlling Tylenchus Semipenetrans Cobb in Vitro and in Vivo in Citrus. Journal of Applied Sciences Research, 8, 5415-5424.

20. Gazim, Z., Moraes, C., Regina, S., Estivaletti, T. and Garcia, D. (2007) Antifungal Activity of the Essential Oil from Calendula officinalis L. (Asteraceae) Growing in Brazil. Brazilian Journal of Microbiology, 39, 61-63.

21. Acosta, L. (2006) Exploración de las propiedades antimicrobianas de extractos vegetales a partir de Calendula officinalis y Tropaeolummajus y su uso potencial en la industria de alimentos y cosmética. Trabajo de grado Uni- versidad de La Sabana, Facultad de Ingeniería, Bogotá.

22. Martínez, J.A. (2012) Natural Fungicides Obtained from Plants, Fungicides for Plant and Animal Diseases. In: Dhanasekaran, D., Ed., Fungicides for Plant Animal Diseases, InTech.http://www.intechopen.com/books

23. Wedge, D., Galindo, J. and Macías, F. (1999) Fungicidal Activity of Natural and Synthetic Sesquiterpene Lactone Analogs. Phyotochemistry, 53, 747-757.

24. Sonoda, R.M. (1977) Effect of Benomyl Drenches on Fusarium Crown Rot of Tomatões. Agricultural Research Center, Port Pierce, 1, 1-5.

25. Amini, J. and Sidovich, D. (2010) The Effects of Fungicides on Fusariumoxysporumf. sp. lycopersicy Associated with Fusarium Wilt of Tomato. Journal of Plant Protection Research, 50, 172-178.

26. Navarrete, R. and Acosta, R. (1999) Reacción de variedades de fríjol común a Fusarium spp. y Rhizoctoniasolani en el altiplano de México. Agronomia Mesoamericana, 10, 37-46. [Paper reference:1]

27. Nel, B., Steinberg, C., Labuschagne, N. and Viljoen, A. (2007) Evaluation of Fungicides and Sterilants for Potential Application Management of Fusarium Wilt of Banana. Crop Protection, 26, 697-705.

28. Silva, O.C., Santos, H.A.A., Dalla Pia, M. and May-De Mio, L.L. (2011) Potassium Phosphite for Control of Downy Mildew of Soybean. Crop Protection, 30, 598-604.

29. Lobato, M.C., Olivieri, F.P., Daleo, G.R. and Andreu, A.B. (2010) Antimicrobialactivity of Phosphites against Dif- ferent Potato Pathogens. Journal of Plant Diseases and Protection, 117, 102-109.

30. Mogollón, A. and Castaño, J. (2011) Efecto de inductores de resistencia en plántulas de plátanodominicohartón (Musa balbisiana AAB) contra Mycosphaerella spp. Revista de la Academia Colombiana de Ciencias Exactas, Físicas y Naturales, 35, 463-471.

31. Walters, D.R., Ratsep, J. and Havis, N.D. (2013) Controlling Crop Diseases Using Induced Resistance: Challenges for the Future. Journal of Experimental Botany, 64, 1263-1280.

32. Hofgaard, I., Ergon, A., Henriksen, B. and Tronsmo, A. (2010) The Effect of Potential Resistance Inducers on Deve- lopment of Microdochium majus and Fusarium culmorum in Winter Wheat. European Journal of Plant Pathology, 128, 269-281.

33. French, R. and Teddy, T. (1986) Métodos de Investigación Fitopatológica. IICA Costa Rica, 158-161.

34. Barnett, L.H. (1998) Illustrated Genera of Imperfect Fungi. The American Phytopathological Society, St. Paul.

35. Leslie, J.F. and Summerell, A. (2006) The Fusarium Laboratory Manual. Blackwell Publishimg, Ames.

36. Cifuentes, J. (2000) Evaluación de la capacidad biocontroladora del hongo Trichodermahazianum cepa nativa Queule sobre Fusarium solani en tomate (Lycopersiconesculentum Mill). http://www.bionativa.cl/pdf/tesis/trichonativa/t13.pdf [P aper reference:1]

37. Muthomi, J., Riungi, W. and Wagacha, J. (2008) Management of Fusarium Head Blight of Wheat Using Antagonistic Microorganisms.http://www.tropentag.de/2007/abstracts/full/448.pdf

38. Stefanova, M., Leiva, A., Larrinaga, L. and Coronado, M. (1999) Actividad metabólica de cepas de Trichodermaspp para el control de hongos fitopatógenos del suelo. Revista de la Facultad de Agronomía de la Universidad del Zulia, 16, 509-516.

39. Selosse, M., Baudoin, E. and Vandenkoornhuyse, P. (2004) Symbiotic Microorganisms, a Key for Ecological Success and Protection of Plants. Comptes Rendus Biologies, 327, 639-648.

40. Santos, F., Souza, P., Resende, M., Pozza, E., Miranda, J., Ribeiro, P. and Manerba, F. (2007) Efeito de extratos ve- getais no progresso de doenças foliares do cafeeiro orgânico. Fitopatologia Brasileira, 32, 59-63.

41. Zapata, R., Sanabria, M. and Rodríguez, D. (2003) Reducción del desarrollo de hongos fitopatógenos con extracto de- cardón lefaria (Cereusdeficiens). Interciencia, 28, 302-308.

42. Hernández, A., Bautista, S. and Velásquez, M. (2007) Prospectiva de extractos vegetales para controlar

enfermedades postcosecha hortofrutícolas. Revista Fitotecnia Mexicana, 30, 119-123.

43. Avis, T., Gravel, V., Antoun, H. and Tweddell, R. (2008) Multifaceted Beneficial Effects of Rhizosphere Microorganisms on Plant Health and Productivity. Soil Biology & Biochemistry, 40, 1733-1740.

44. Monsalve, J., Viteri, S., Rubio, N. and Tovar, F. (2012) Efecto del fosfito de potasio en combinación con el fungicida me- talaxyl + mancozeb en el control de mildeo velloso (Peronospora destructor Berk) en cebolla de bulbo (Allium cepa L.). Revista Facultad de Agronomía Universidad Nacional de Medellín, 65, 6317-6325.

Chapter 10

Localization of Probe-Accessible Chitin and Characterization of Genes Encoding Chitin-Binding Domains During Rice–Magnaporthe Oryzae Interactions

Susumu Mochizuki • Ken-ichiro Saitoh • Eiichi Minami • Yoko Nishizawa

[1]*Plant–Microbe Interactions Research Unit, Division of Plant Sciences National Institute of Agrobiological Sciences TsukubaJapan*

[2]*Advanced Career Development Center for Agro-innovation Tokyo University of Agriculture and Technology TokyoJapan*

ABSTRACT

We examined the localization of probe-accessible chitin in *Magnaporthe oryzae*, which causes rice blast disease, during the early infection process and the functions of two genes encoding a chitin-binding domain (ChBD). Invasive hyphae in the first-invaded rice cell showed little staining with fluorescently labeled wheat germ agglutinin, a probe to detect chitin. However, in the second-invaded cell, hyphae showed strong fluorescence, and treatment with chitinase diminished the signal. Fourteen *ChBD* genes encoding family 18 carbohydrate-binding module (CBM18) were isolated from a Japanese strain of *M. oryzae*, Ina86-137. Reverse transcription-polymerase chain reaction analysis demonstrated that *ChBD-1, ChBD-6, ChBD-8, ChBD-13,* and*ChBD-15* are expressed in the rice sheath. Gene-targeted disruptants of *ChBD-1* and *ChBD-15*had no significant differences in invasive growth, pathogenicity, or tolerance to chitinase compared to the wild type. These results suggest that *M. oryzae* has a mechanism to evade being a substrate for the host chitinase in the first-invaded cell, but neither *ChBD-1* nor *ChBD-15* contributes to this mechanism.

Keywords: CBM18 Chitin-binding domain Chitin elicitor Fungal cell wall *Magnapor the oryzae* Rice

INTRODUCTION

Materials and methods

Fungal isolates, rice plants, and culture conditions

Magnaporthe oryzae strains Ina86-137 (MAFF Gene Bank stock number MAFF101511) and TH68-126 (MAFF101517), Japanese pathogenic isolates from rice, were cultured on potato dextrose agar (PDA; Difco Laboratories, Detroit, MI, USA) at 25°C. To collect conidia, the fungi were cultured on oatmeal agar medium (5% oatmeal powder, 1.5% agar) for 7 days at 25°C in darkness. Production of conidia was induced by irradiation with a blacklight blue lamp for 4 days after the aerial mycelia were removed using a sterilized brush. Rice plants (*Oryza sativa*L. cv. Nipponbare BL no. 2, compatible with Ina86-137, and cv. Nipponbare BL no. 5, compatible with TH68-126) were grown hydroponically for 3–4 weeks in a nutrient solution as described previously (Tanabe et al. 2006), using a growth chamber with a 14-h light (28°C, 5,000 lx)/10-h dark (25°C) cycle.

Chitin staining

WGA-Alexa Fluor 488 conjugate (Molecular Probes, Eugene, OR, USA) was used according to Ramonell et al. (2005) to stain chitin in the infection structures of *M. oryzae* that formed on rice leaf sheaths. Trimmed leaf sheaths were washed in 1× phosphate-buffered saline (PBS; pH 7.4) containing 0.1% Triton X-100 for 20 min and stained with WGA-Alexa Fluor 488 at 10 μg/mL in 1× PBS (pH 7.4) containing 0.1% Triton X-100 for 4 h at 4°C. The sheaths were then washed in the PBS buffer for 3 h and in fresh buffer overnight. Propidium iodide (PI; 50 μg/mL) in 1× PBS (pH 7.4) was used as a counterstain. For enzymatic digestion of chitin, the inoculated rice leaf sheath was incubated with 0.5 U/mL of purified chitinase from *Trichoderma viride* (Sigma, St. Louis, MO, USA) in 1× PBS (pH 7.4) with 1% Tween 20 for 18 h at 25°C, then washed with 1× PBS (pH 7.4)–0.1% Triton X-100 three times before staining with WGA-Alexa Fluor 488.

Confocal microscopy

Confocal microscopy was performed using a TCS SP5 instrument (Leica, Wetzlar, Germany). Fluorescence was excited with an argon laser at 488 nm and detected at wavelengths of 500–520 nm [WGA-Alexa Fluor 488 and green fluorescent protein (GFP)] or 600–620 nm (PI). Images were processed and arranged using LAS AF software (Leica) and Adobe (San Jose, CA, USA) Photoshop CS4 software.

RT-PCR assay

Total RNA was extracted from mycelia growing in YG broth (0.5% yeast extract, 2% glucose) and from rice leaf sheaths at 40 h post-inoculation (hpi) using an RNeasy Mini Kit (Qiagen, Valencia, CA, USA). The mycelia and infected rice sheaths were placed in screw-capped microtubes containing 450 μL of the kit's RLT buffer and five zirconia beads (2 mm diameter), then homogenized using a Micro Smash MS-100 system (Tomy Seiko, Tokyo, Japan) by shaking the tubes three times (30 s each, at 4,000 rpm). The extracted total RNA (500 ng) was used as the template for a reverse transcription reaction using a ReverTra Ace α Kit (Toyobo, Osaka, Japan) in a 20-μL reaction mixture. The mixture was diluted with 100 μL of sterile distilled water, and 3-μL samples were used for the PCR, which was performed with each of the primer sets listed in Table 1 at 400 nM and with 5% dimethyl sulfoxide, using a GoTaq kit (Promega, Madison, WI, USA). The initial denaturation step was 2 min at 95°C, followed by 30 (mycelia) or 35 (infected rice) cycles of amplification consisting of 30 s each at 95, 60, and 72°C.

Vectors and *Agrobacterium*-mediated transformation

Vectors used to knockout *ChBD* genes and for *Agrobacterium*-mediated transformation of *M. oryzae* were constructed using the GFP-tagging gene knockout (GGKO) method of Saitoh et al. (2008). Table 1 lists the primers used to construct the knockout vectors. Fungal DNA for PCR analysis was prepared using a simple method developed by Saitoh et al. (2006), and gene disruption was confirmed by PCR using the primers listed in Table 1.

Pathogenicity test

For the spray inoculation, 2-week-old rice seedlings grown in a greenhouse were sprayed with conidial suspensions (5×10^5 conidia/mL) of *M. oryzae* in 0.05% Triton X-100 using the method of Nishizawa et al. (1999). The inoculated seedlings were incubated in a humid chamber at 25°C for 20 h, then cultivated in a greenhouse. At least three independent mutant lines per target gene were analyzed, with the test carried out twice.

Table 1. Primers used in this study

Primer	Sequence	Target gene
(1) RT-PCR analysis		
01RT-U[a]	AGGCCTCAAGTGCCCTGGA	*ChBD-1*
01RT-L[a]	CCGCCGCAGTCGATGAAGA	*ChBD-1*
02RT-U	CGTGCCAGGCGTATTGCGA	*ChBD-2*
02RT-L	AAACGAGGTCGACGTGCCC	*ChBD-2*
03RT-U	ATACTGCGGCGTGGGATGC	*ChBD-3*
03RT-L	GCCATAGTCAGCCTGGCAG	*ChBD-3*
04RT-U	GGCCACAGTGAGCGTTTCC	*ChBD-4*
04RT-L	GCGATTTCCCTCCGCACCT	*ChBD-4*
06U03	AAAGACTCGGACGGCATCA	*ChBD-6*
06RT-L	CCCCAGTTCTTCTGGCACC	*ChBD-6*
07RT-U	GCCGCTAACCCTGGGCATT	*ChBD-7*
07RT-L	AATCACCGGCACAGCTCCC	*ChBD-7*
08RT-U	TCGGGCGCCTTCACCTTTG	*ChBD-8*
08RT-L	TCGTCCGTCTTGCCGTTGG	*ChBD-8*
09U04	GTGTTGCGTTTGGCACACA	*ChBD-9*
09L03	ACCATGGCTTTGCTGTCGT	*ChBD-9*
10RT-U	TGCCCGGGTGAACTTGGGA	*ChBD-10*
10RT-L	ACCCACTGATCCGTGTCCC	*ChBD-10*
11RT-U	TCCGGCTTCGGCAACTGCT	*ChBD-11*
11L06	CCCCCCCAACTGTCTATTTT	*ChBD-11*
12RT-U2	TGCCGAGTTTTGGTACCGT	*ChBD-12*
12RT-L2	TCCCCAATGCTTATCGCTG	*ChBD-12*
13RT-U	ACAGACCGGCAGCTACGAC	*ChBD-13*
13RT-L	TGTCGAGGCTGGTGAAGCC	*ChBD-13*
14RT-U	GAAGCTGCGGTGGCAAGGA	*ChBD-14*
14RT-L	GATGGAGTGGCGGCTGCAT	*ChBD-14*
15RT-U3	AACAACGCCACCTGCCAGG	*ChBD-15*
15RT-L[a]	TGGCGTCGGCGGTGGACTT	*ChBD-15*
ActQ1[b]	TCTTCGAGACCTTCAACGCC	*actin*
ActQ2[b]	ACCGGAGTCGAGCACGATAC	*actin*
(2) Construction of knockout vectors[c]		
01KO5U	ATGCCTGCAGATGGCATGGGATCGCAGGA	*ChBD-1*
01KO5L	ATGCGGATCCTTTTTGGGGGGGATTTTGTG	*ChBD-1*
01KO3U	ATGCCCGCGGAAAATGCGGGGTGCGTTTG	*ChBD-1*
01KO3L	ATGCGGTACCGGATGGAATCAAGCTCAACG	*ChBD-1*
15KO5U	ATGCGTCGACTTTGGTATTGTATACCCGCG	*ChBD-15*
15KO5L	ATGCGGATCCTTTTGGTTGGGCTTGGGGTT	*ChBD-15*
15KO3U	ATGCCCGCGGAAACAAAGGGTGTCACCCCA	*ChBD-15*
15KO3L	ATGCGGTACCGGCTGATGCTTCCTACCTTA	*ChBD-15*
(3) Confirmation of the gene knockout		
HPT5-1	ATGAAAAAGCCTGAACTCACCGCGA	*HPT*
HPT3-1	TCCATCACAGTTTGCCAGTGATACA	*HPT*
01U01	CTCTGCCAAGCCACAGTTA	*ChBD-1*
01L01	GGTTTGGGAGAGAAGAGAG	*ChBD-1*
15U01	TTTCCCGCACGCTACTCTA	*ChBD-15*
15L01	TCACGAGTGGGTATAGGCA	*ChBD-15*
15RT-U	CCATCGTCCTCCTCACCCT	*ChBD-15*

[a]These primers were also used in identification of the knockout mutants
[b]Primers reported previously (Li et al. 2007)
[c]Underlines indicate the restriction endonuclease recognition site

For the evaluation of degree of hyphal growth, the midrib area of detached rice sheaths was inoculated with a conidial suspension (10^6 conidia/mL, 20 hpi, or 10^5 conidia/mL, 48 hpi) as described previously (Tanabe et al. 2006). After a 20 or 48 h incubation at 25°C in darkness, the inoculated sheath was hand-trimmed for microscopic observation. The degree of invasion for all appressoria on each 1.5-cm-long leaf sheath at 20 hpi and for ca. 150 intact appressoria at 48 hpi was assessed as "multicell invasion," "single-cell invasion," or "no invasion," depending on whether appressoria had penetrated more than one rice cell, one cell, or no cells, respectively (Tanabe et al. 2006). Three to five leaf sheaths per line were observed. For assaying the sensitivity of germ tubes to chitinase, 20 μL of a conidial suspension (5.0×10^4 conidia/mL) was incubated on plastic coverslips for 16 h at 25°C in darkness. Appressorium-forming conidia were incubated with 1× PBS buffer with 0.5% Tween 20 containing 0.5 U/mL chitinase from *T. viride* for 8 h. Statistical analyses (Tukey's test and Dunnett's test) were performed using JMP8 software (SAS Institute, Cary, NC, USA).

RESULTS

Localization of probe-accessible chitin in infection structures

To determine whether the surface components of the rice blast fungus change after invasion of rice cells, rice leaf sheaths were inoculated with a conidial suspension of a virulent isolate and stained with WGA-Alexa Fluor 488 probe, a chitin-binding lectin conjugated with a green fluorescent dye. The stained sheath sections were then observed using confocal laser microscopy. Surfaces of the germ tube and appressorium fluoresced strongly. During early infectious growth up to 30 hpi (Fig. 1a), the primary and invasive hyphae tended to be stained. However, the intensity of the fluorescence of the invasive hyphae inside the first-invaded epidermal cell diminished after 30 hpi. In contrast, once the invasive hyphae entered the second-invaded cell,

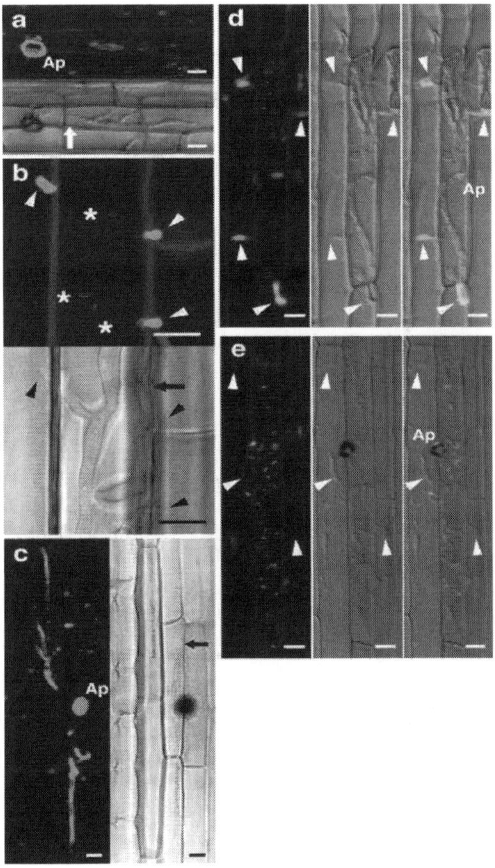

Figure 1. Wheat germ agglutinin (WGA) probe binding to the rice blast fungus growing in epidermal cells of rice leaf sheath. **a–c** Inoculated, detached rice sheaths were treated with WGA-Alexa Fluor 488 (green channel) and propidium iodide (PI) (red channel, staining nuclei and plant cell wall) at 30 hpi (**a**), 36 hpi (**b**), and 42 hpi (**c**). **d**, **e** Effect of pretreatment with chitinase on WGA binding. Samples were stained with WGA-Alexa Fluor 488 and PI after treatment without (**d**) and with chitinase (**e**) at 36 hpi. The *fluorescence-field panels* show superimposed confocal images in which stacked Z-series images corresponding to the appressoria and epidermis obtained in the green channel and an image in the red channel were merged. The *bright-field panels* are differential interference contrast images. In the rightmost panels in **d** and **e**, the fluorescence image and the bright field image are superimposed.*Arrows* indicate first-invaded epidermal cells. *Arrowheads* denote hyphae that had just invaded second cells. *Asterisks* show WGA-stained septa. *Ap* appressorium. *Bar* 10 μm

they fluoresced strongly (Fig. 1b), and the fluorescent signals lasted for a longer time (Fig. 1c). Focal plane scanning confirmed that the hyphae were inside the epidermal cells. The test was performed using two combinations of compatible interactions (rice cv. Nipponbare BL no. 2 and blast strain Ina86-137, or cv. Nipponbare BL no. 5 and TH68-126), and similar staining patterns were observed for both combinations.

We next confirmed whether WGA-Alexa Fluor 488 probe stained chitin on *M. oryzae*. For treatment with chitinase, an 18-h incubation in buffer containing 1% Tween 20 was required prior to WGA staining. Figure 1d shows that hyphae in the second-invaded cells were successfully stained in this buffer without chitinase. In comparison, pretreatment of the inoculated rice sheaths with chitinase diminished the WGA staining of the invasive hyphae in the second-invaded cells (Fig. 1e), strongly suggesting that we had observed WGA probe-accessible chitin in the infection structures of *M. oryzae*.

Structure of genes encoding chitin-binding domains

Carbohydrate-binding modules (CBMs) are classified into 54 families based on sequence similarity and protein folds (http://www.cazy.org/; Cantarel et al. 2009). Several of these modules, such as CBM5, 12, 14, and 18, possess chitin-binding activity. A search of the *M. oryzae* genome sequence database (ver. 4; Magnaporthe Sequencing Project. R. Dean, Fungal Genomics Laboratory, North Carolina State University, http://www.fungalgenomics.ncsu.edu; and MIT/Harvard Broad Institute, http://www.broad.mit.edu) revealed 15 genes encoding CBM18 motifs (designated *ChBD-1–ChBD-15*; Table 2), including seven lectin-type *ChBD* genes (*ChBD-1–ChBD-6, ChBD-15*), one class III chitinase gene (*ChBD-7*), three class V chitinase genes (*ChBD-8–ChBD-10*), and two chitin deacetylase genes (*ChBD-11* and *ChBD-12*). Because all 15 proteins possess a signal sequence, they appear to be secreted. The ChBD found in the Avr4 protein of *C. fulvum* belongs to the CBM14 family (van den Burg et al. 2003), but we found no genes encoding CBM14, CBM5, or CBM12 in the *M. oryzae* database.

Table 2. Summary of the structural analysis of *ChDB* genes

Gene name	Gene locus	Predicted function	Deduced protein length	Differences in amino acid sequence compared with 70-15
ChBD-1	MG05351.5	Chitin binding protein	479	An insertion of 58 residues
ChBD-2	MG00227.5	Chitin binding protein	396	None
ChBD-3	MG08729.5	Chitin binding protein	266	None
ChBD-4	MG02142.5	Chitin binding protein	1,268	An insertion of 87 residues
ChBD-5[a]	MG03784.5	Chitin binding protein	–	–
ChBD-6	MG06771.5	Chitin binding protein	224	None
ChBD-7	MG03599.5	Chitinase	422	None
ChBD-8	MG06594.5	Chitinase	428	3 substitutions: T404S, D416V, N423K
ChBD-9	MG01336.5	Chitinase	476	None
ChBD-10	MG04534.5	Chitinase	531	None
ChBD-11	MG09159.5	Chitin deacetylase	529	4 substitutions: A86V, I100V, K535Q, F551S and a deletion of 39 residues
ChBD-12	MG05023.5	Chitin deacetylase	352	None
ChBD-13	MG09717.5	Dehydrogenase	718	Substitution: P211L
ChBD-14	MG10431.5	*β*-Glucanase	793	None
ChBD-15	MG09248.5	Chitin binding protein	102	None

[a] *ChBD-5* was not obtained from Ina86-137

We compared the genomic sequences corresponding to the 15 *ChBD* genes in a Japanese isolate, Ina86-137, with the sequence of experimental strain 70-15 (Dean et al. 2005). *ChBD-5* from Ina86-137 was not amplified, even though the gene was successfully isolated from another *M. oryzae* strain, Guy11. The nucleotide sequences of the other 14 *ChBD* genes obtained from Ina86-137 have been registered in DDBJ under accession numbers AB513115–AB513128. Figure 2a shows their putative protein structures. As summarized in Table 2, *ChBD-1* of Ina86-137 carries a 174-bp insert, resulting in the introduction of another CBM18 motif, and *ChBD-4*and *ChBD-11* have a 261-bp insert and a 117-bp deletion, respectively. The sequences of open reading frames, introns, and adjacent regions up to 500–1,900 bp of the other genes, which were confirmed to produce knockout mutants, showed more than 99% identity. All amino acid sequences of CBM18, except those of ChBD-1 and ChBD-11, were

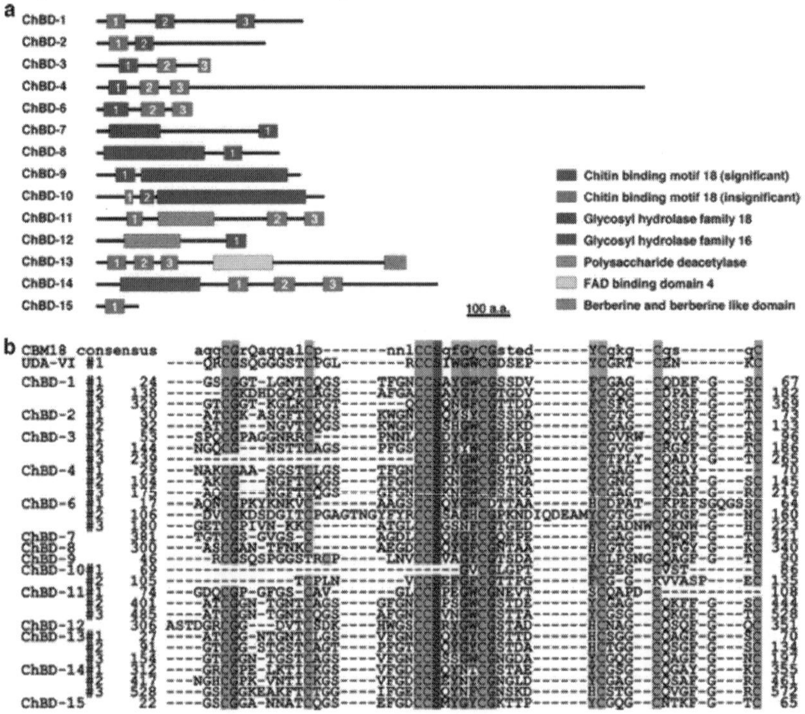

Figure 2. Position and sequence of CBM18 in *ChBD* genes. **a** Schematic of structures of proteins encoded by the*ChBD-1–ChBD-15* genes in *M. oryzae* Ina86-137. Conserved domains detected by Pfam analysis (http://pfam.sanger.ac.uk/search) are color-coded. Each CBM18 is numbered beginning at the N-terminus. **b** Comparison of CBM18 amino acid sequences found in ChBD-1–ChBD-15. The amino acid sequences were aligned with the CBM18 consensus sequence and the CBM18 of *Urtica dioica* isolectin. The triad of aromatic residues and the serine residue are indicated by *yellow* and *red boxes*, respectively. Conserved cysteines and glycines are indicated by *green* and *blue boxes*, respectively

conserved between strains 70-15 and Ina86-137. Figure 2b shows an alignment of the CBM18 amino acid sequences found in ChBD-1–ChBD-15, except for ChBD-5, along with the CBM18 consensus sequence and the sequence for CBM18 of *Urtica dioica* isolectin, for which the crystal structure and amino acid residues participating in chitin binding have been determined (Harata and Muraki 2000). CBM18 contains several disulfide bonds, and the corresponding cysteine residues, as well as adjoining glycine residues, are well conserved in the ChBD-1–ChBD-15 proteins. The central region of CBM18, which contains three aromatic amino acid residues and one serine residue that participate in chitin binding (Harata and Muraki 2000), is especially highly conserved.

Expression analysis of *ChBD* genes using RT-PCR

To analyze the expression of the *ChBD* genes at the infection stage, we conducted RT-PCR using primer sets designed to specifically amplify each of the 14 *ChBD* genes. Five ChBD genes, *ChBD-1*, *ChBD-6*, *ChBD-8*, *ChBD-13*, and *ChBD-15*, were expressed in the rice sheath at 40 hpi, but only *ChBD-8* was expressed in vegetatively growing mycelia (Fig. 3). No expression of other*ChBD* genes was detected at the stages tested. These genes may be expressed under specific conditions or may be pseudogenes.

Knockout of *ChBD* genes

We selected *ChBD-1* and *ChBD-15* from the three lectin-type *ChBD* genes expressed in the inoculated rice sheath and generated gene-targeted knockout mutants of each using the GGKO method of Saitoh et al. (2008). Successful gene-targeted disruption was confirmed by PCR (Fig. 4). No significant differences in radial growth on PDA medium up to 10 days or sporulation on oatmeal agar

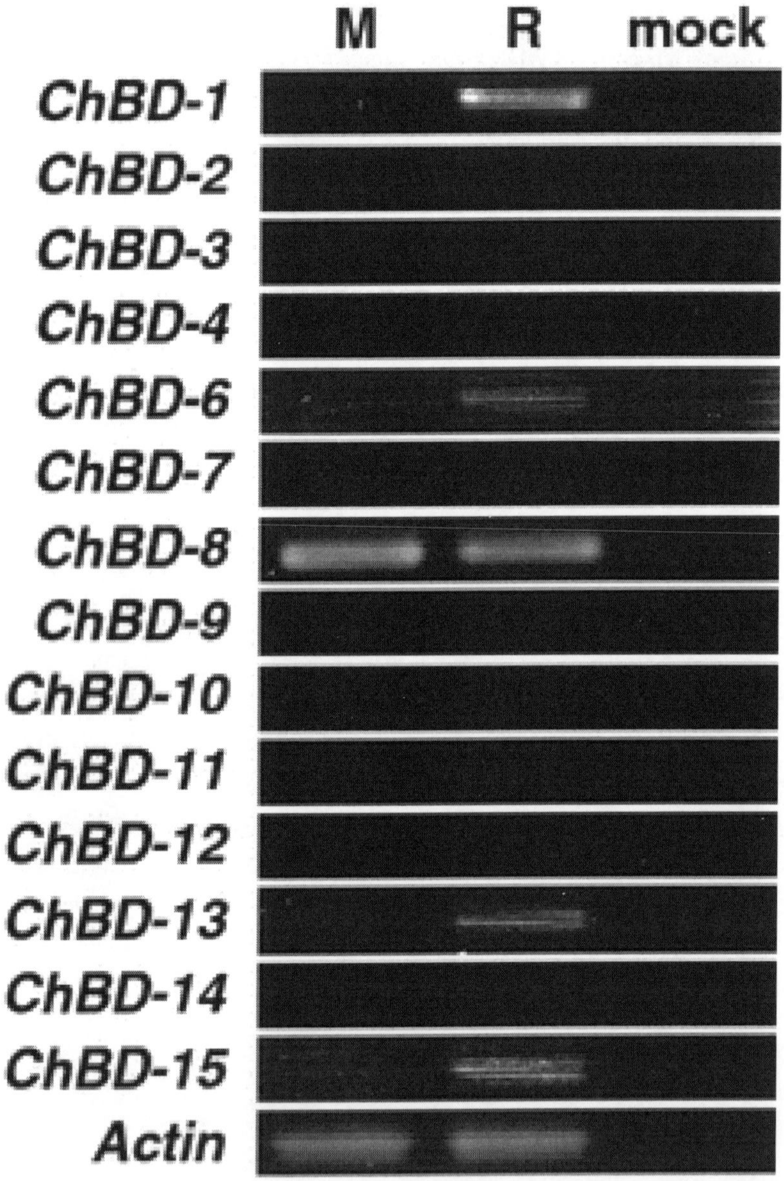

Figure 3. Expression analysis of *ChBD* genes using RT-PCR. RNA from mycelia (*M*) grown in yeast glucose medium and rice sheath (*R*) at 40 hpi were used as templates. *Mock* indicates results obtained using RNA extracted from uninoculated leaf sheaths

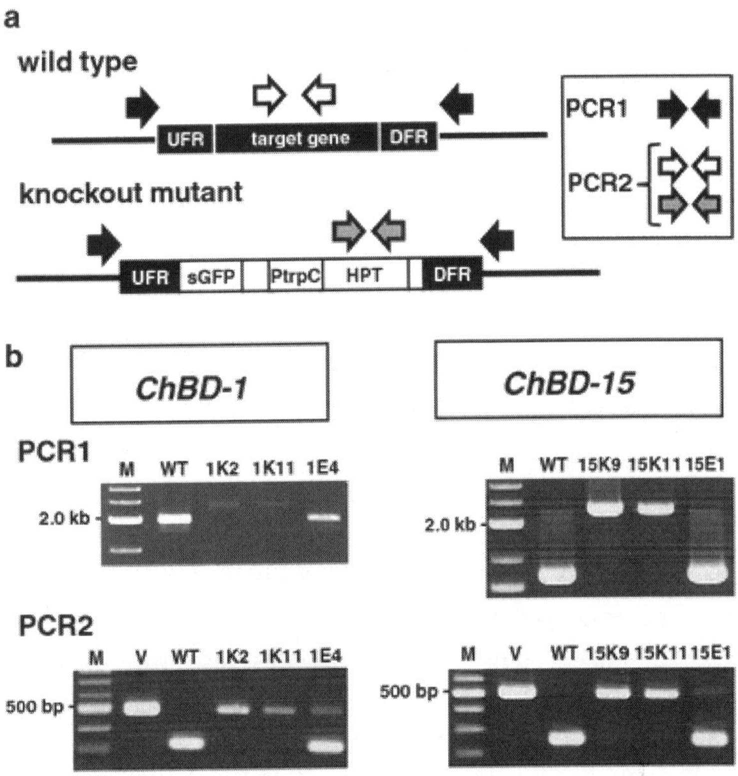

Figure 4. Confirmation of knockout mutants by genomic PCR. **a** Schematic representation of two PCR sets to confirm gene disruption. PCR1 was performed with primers corresponding to areas outside the upstream flanking regions (UFRs) and downstream flanking regions (DFRs; *solid arrows*). Knockout mutants were expected to generate a single band in PCR1 that differed in size from that of the wild-type Ina86-137 and ectopic transformants. PCR2 consisted of multiplex PCR using primers specific for the target gene (*open arrows*) and the hygromycin phosphatase gene (*HPT*; *gray arrows*). The HPT fragment should be about 500 bp. In PCR2, the wild type was expected to amplify the target gene, whereas the knockout mutants were expected to amplify HPT and a disrupted target gene fragment from the ectopic transformants, thus generating products from both genes. **b** Results of both PCR sets. In the names of the isolates, "*K*" indicates a knockout mutant, and "*E*" indicates an ectopic transformant. In PCR2, the binary vector DNA for the target gene disruption was also used (*V*). *M* DNA size marker

were detected between the two knockout mutants and wild-type controls ($n = 3$; Tukey's test and Dunnett's test). Appressorium formation was examined on cover glasses made of plastic or glass, on cellophane, and on plant epidermis samples, and no morphological aberrations or reductions in frequency of appressoria formation were observed.

Expression analysis of selected *ChBD* genes by GFP fluorescence

Because the gene-disrupted lines possessed the GFP gene under the control of their own promoter, we were able to observe stage- and site-specific expression of each gene using GFP fluorescence. Neither *ChBD-1* nor *ChBD-15* was expressed before appressoria formed. During the appressorium formation and intracellular invasive growth, both *ChBD-1* and *ChBD-15* were expressed in appressoria (Fig. 5), suggesting that the two genes function before penetration.*ChBD-1*, but not *ChBD-15*, was also expressed in the primary hyphae (16 hpi). Neither of the genes was expressed in expanded invasive hyphae (36 hpi), suggesting that they function during an early stage of infection.

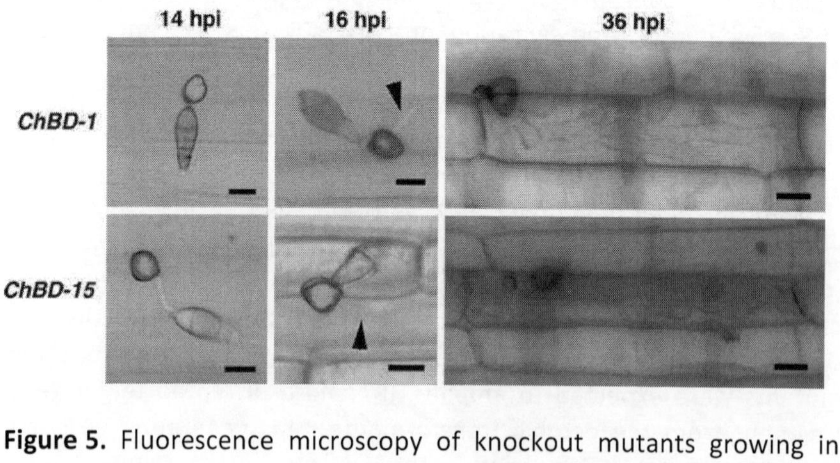

Figure 5. Fluorescence microscopy of knockout mutants growing in rice leaf sheaths. All panels are superimposed images of the stacked Z-series confocal GFP fluorescent images corresponding to the appressoria and epidermis and a differential interference contrast image. *Arrowheads* show primary hyphae. *Bar* 10 μm

Pathogenicity and tolerance to chitinase of Δ*ChBD-1* and Δ*ChBD-15*

To analyze the pathogenicity of the *ChBD-1* and *ChBD-15* knockout mutants (Δ*ChBD-1* and Δ*ChBD-15*) after spray inoculation, we compared lesion areas at 8 days post-inoculation (dpi). Typical susceptible lesions developed in the knockout lines (Fig. 6a), and no significant differences were observed in the degree of lesion formation among them.

We next evaluated infection ability more precisely with a leaf sheath assay. Rice leaf sheaths were inoculated with a conidial suspension of each knockout mutant (two lines per disruptant), and the extent of elongation of the infection hyphae was assessed at 20 and 48 hpi. No significant differences were observed between mutants and the wild type (Fig. 6b, c). No morphological alterations during germination and appressoria formation on the rice sheath were detected.

To examine the effect of gene disruption of *ChBD-1* or *ChBD-15* on tolerance to chitinase, appressorium-forming conidia of Δ*ChBD-1* and Δ*ChBD-15* were treated with chitinase and the percentage of lysis of germ tubes was assessed. As shown in Fig. 6d, ca. 50% of the germ tubes were lysed, which was not significantly different from the percentage for the wild-type strain.

DISCUSSION

Chitin and its fragments, *N*-acetylchitooligosaccharides, are representative microbe-associated molecular patterns of fungi that trigger various defense responses in plants at subnanomolar levels (Shibuya and Minami 2001). Thus, pathogenic fungi are very likely to have mechanisms to avoid generating the chitin elicitor during the infection process. Because *M. oryzae* possesses more genes encoding putative chitin-binding proteins than do nonphytopathogenic fungi (Dean et al. 2005), we speculated that *M. oryzae* has a mechanism to

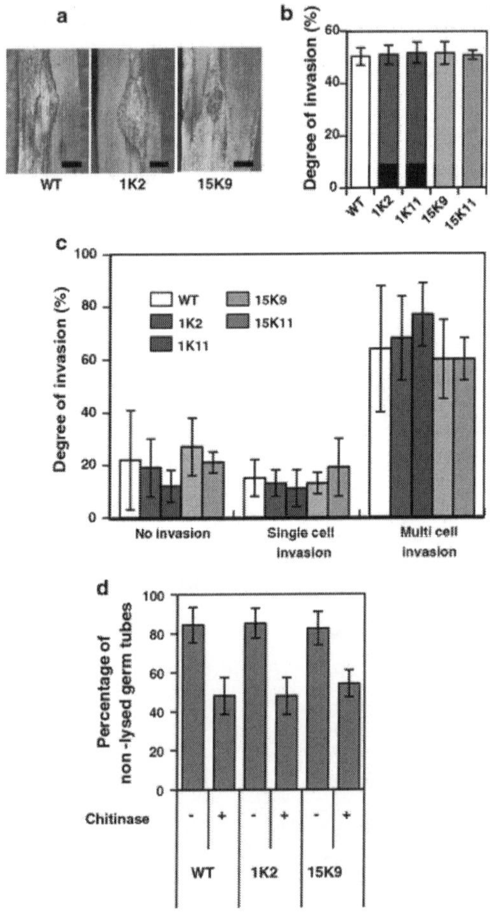

Figure 6. Evaluation of pathogenicity of *Magnaporthe oryzae* and rice cell wall sensitivity to chitinase of *ChBD*knockout mutants. **a** Lesions on rice leaf blade at 8 dpi. Two-week-old rice seedlings (16–20 plants per mutant line) were sprayed with a conidial suspension of each mutant. Inoculation tests were performed twice, and representative lesions are shown. *Bar* 1 mm. **b** Degree of hyphal penetration into rice leaf sheath epidermal cells at 20 hpi. **c** Degree of hyphal growth in rice leaf sheath epidermal cells at 48 hpi. **d** Percentage of nonlysed germ tubes after 8 h incubation in reaction buffer without and with chitinase. Mutant line names correspond to those in Fig. 4. *Bars* indicate standard deviation. Statistical analyses (Tukey's test and Dunnett's test) demonstrated no significant difference between wild-type controls and mutants

prevent being a substrate of rice chitinases and that chitin-binding proteins may contribute to this mechanism.

In the first-invaded rice cell, invasive hyphae had little fluorescence from fluorescently labeled WGA, consistent with the results reported by Fujikawa et al. (2009). Our present study demonstrated that when the hyphae expanded to neighboring cells, they fluoresced strongly. Because invasive hyphae, which develop from primary hyphae, are sealed within the plant membrane, designated an extrainvasive hyphal membrane (EIHM) (Kankanala et al. 2007), access of the WGA probe to the cell walls of the invasive hyphae could be restricted. However, the septae of invasive hyphae inside the first-invaded epidermal cell fluoresced (Fig. 1b), indicating that the WGA probe had accessed hyphae within the rice cell. Invasive hyphae growing in neighboring cells are also sheathed in EIHM (Kankanala et al. 2007); therefore, the accessibility of WGA probe to fungal cell walls should be similar in first- and second-invaded rice cells. Pretreatment with chitinase diminished WGA binding to the hyphae in the second-invaded cells (Fig. 1e), indicating that the observed fluorescence resulted from the binding of the WGA probe to chitin. Thus, our data strongly suggest that chitin in the invasive hyphae within the first-invaded epidermal cell is masked or modified, whereas chitin in the hyphae in neighboring cells is accessible to the WGA probe. These observations imply that the rice blast fungus has a mechanism to protect the cell wall against host chitinase and prevent the release of chitin elicitor in the early infection stages, as suggested for other host–pathogen interactions (El Gueddari et al. 2002; van den Burg et al. 2006). In contrast, hyphae in second- and later-invaded cells may not need to limit chitin exposure to be virulent. Invasive hyphae in second- and later-invaded rice cells have been reported to differ in behavior and often in appearance from those in first-invaded cells (Kankanala et al. 2007). Our results indicate that the cell wall properties of invasive hyphae in the first-invaded rice cell also differ from those in the second- and later-invaded rice cells, which implies that host–pathogen interactions in neighboring cells also differ from those in the first-invaded cell. Kishimoto et al. (2010) reported that transgenic rice plants with enhancement of chitin elicitor signaling suppress multi-cell invasion of *M. oryzae,* but do not influence the degree of one-cell invasion, which probably reflects the distribution of chitin in the *M. oryzae* cell wall.

We found 15 *ChBD* genes encoding CBM18 motifs in the genome sequence database, of which 14 were successfully cloned from a Japanese isolate. RT-PCR analysis of the 14 genes indicated that *ChBD-1, ChBD-6, ChBD-8, ChBD-13,* and *ChBD-15* are expressed in the rice sheath. We produced gene-disrupted mutants of the two lectin-type *ChBD* genes using the GGKO method (Saitoh et al. 2008), which allows observation of stage- and site-specific expression of the target gene using GFP fluorescence. Unlike RT-PCR analysis, GFP fluorescence observations revealed that only *ChBD-1* is expressed in rice cells. This inconsistency is likely because the infection process of each conidium proceeds asynchronously, and the RNA sample was obtained from tissue containing fungi at different stages of infection. Thus, live-cell imaging of the mutants generated by the GGKO method is advantageous in analyzing stage- and site-specific gene expression.

Expression data suggest that *ChBD-1* and *ChBD-15* function in the penetration stage. However, neither Δ*ChBD-1* nor Δ*ChBD-15* showed defects in appressoria formation, invasive growth, pathogenicity, or tolerance to chitinase. The staining pattern of fluorescently labeled WGA at 30, 36 and 42 hpi also did not differ from that of the wild-type strain (data not shown). Although these *ChBD* genes might have functional redundancy, our data strongly suggest that neither gene contributes to pathogenicity or protection from chitinase, which implies that the participation of chitin-binding proteins with CBM18 in virulence is low in *M. oryzae*.

A secretory protein, Ecp6, which harbors three lysin motifs (LysMs), was identified as a virulence factor of *C. fulvum* (Bolton et al. 2008), and quite recently, Ecp6 was found to sequester chitin oligosaccharides, preventing elicitation of host immunity (de Jonge et al. 2010). LysMs, recognized in prokaryotes and plants as a family of CBMs, are found in receptors for chitin elicitors CEBiP and CERK1 (Kaku et al. 2006; Miya et al. 2007). Putatively secreted LysM-containing proteins are widespread in the fungal kingdom, including *M. oryzae* (Bolton et al.2008). Secretory proteins containing LysM domains rather than CBM18 may play important roles in virulence of *M. oryzae*.

Recent studies have revealed that cell wall chitin is physically masked by other cell wall components. Rappleye et al. (2007) reported the presence of α-1,3-glucan, which masks the trigger of the host immune system, in the outermost layer of the cell wall of *Histoplasma capsulatum*, a pathogenic fungus of mammals. Macrophages recognize the infection by *H. capsulatum* in part through the interaction of β-glucan from the fungal cell wall and the receptor dectin-1, and thus α-1,3-glucan blocks recognition by covering β-glucan (Rappleye et al. 2007). Similarly in *M. oryzae*, Fujikawa et al. (2009) showed that α-1,3-glucan masks both chitin and β-1,3-glucan in the cell walls of *M. oryzae* at 24 hpi. They also demonstrated that the presence of α-1,3-glucan results in increased tolerance to chitinase digestion. Thus, masking chitin by α-1,3-glucan is a major strategy of *M. oryzae* to evade attack by chitinase.

Another possible explanation for the unaltered pathogenicity is that chitin may be modified to avoid hydrolysis by chitinase. One mechanism that protects the fungal cell wall from the host chitinase may be the conversion of chitin to chitosan by chitin deacetylase secreted from the pathogen. For example, El Gueddari et al. (2002) found that chitin was exposed on surfaces of infection structure of the rust fungi *Puccinia graminis* and *Uromyces fabae* that form on plant cuticles, whereas chitosan was exposed on surfaces of the infection hyphae. Deuteromycetes that produce chitin deacetylase are all plant pathogens and secrete chitin deacetylase during plant cell penetration (Hekmat et al. 2003). Determining whether the reduced affinity of WGA for the invasive hyphae of *M. oryzae* is also due to the conversion of chitin to chitosan will be revealing. Fujikawa et al. (2009) have also reported that a major polysaccharide labeled at the accessible surface of *M. oryzae* invasive hyphae is chitosan, in addition to α-1,3-glucan. Of the*ChBD* genes listed in Table 2, *ChBD-11* and *ChBD-12* encode a putative chitin deacetylase. Δ*ChBD-12* showed no defects in invasive growth, pathogenicity or staining pattern of the fluorescently labeled WGA (K. Saitoh, S. Mochizuki, and Y. Nishizawa, unpublished data). However, we were unable to obtain Δ*ChBD-11*, because the mutation may be lethal. Examination of the stage and site of *ChBD-11* expression should provide information on the contribution of chitosan in the protection of the *M. oryzae* cell wall during infection.

ACKNOWLEDGMENTS

We are grateful to Drs. Nagao Hayashi and Shigeru Tanabe at the National Institute of Agrobiological Sciences, Tsukuba, Japan; Dr. Hayashi provided strains of *M. oryzae*(MAFF101511 and MAFF10517) and Dr. Tanabe provided technical advice regarding inoculation. We are also grateful to Dr. Hiroyuki Satoh at the National Institute of Crop Science, Tsukuba, Japan, for providing rice seeds of Nipponbare BL no. 2 and BL no. 5. We also thank Kyoko Iwasaki, Emi Nakajima, and Hiroko Kurano for daily assistance. This research was supported by the program for Promotion of Basic Research Activities for Innovative Biosciences (PROBRAIN). The experiments performed comply with the current laws of Japan.

REFERENCES

1. Bolton MD, van Esse HP, Vossen JH, de Jonge R, Stergiopoulos I, Stulemeijer IJE, van den Berg GCM, Borrás-Hidalgo O, Dekker HL, de Koster CG, de Wit PJGM, Joosten MHAJ, Thomma BPHJ (2008) The novel *Cladosporium fulvum* lysin motif effector Ecp6 is a virulence factor with orthologues in other fungal species. Mol Microbiol 69:119–136

2. Broglie K, Chet I, Holliday M, Cressman R, Biddle P, Knowlton S, Mauvais CJ, Broglie R (1991) Transgenic plants with enhanced resistance to the fungal pathogen *Rhizoctonia solani*. Science 254:1194–1197

3. Cantarel BL, Coutinho PM, Rancurel C, Bernard T, Lombard V, Henrissat B (2009) The Carbohydrate-Active EnZymes database (CAZy): an expert resource for glycogenomics. Nucleic Acids Res 37:D233–D238

4. de Jonge R, van Esse HP, Kombrink A, Shinya T, Desaki Y, Bours R, van der Krol S, Shibuya N, Joosten MHAJ, Thomma BPHJ (2010) Conserved fungal LysM effector Ecp6 prevents chitin-triggered immunity in plants. Science 329:953–955

5. Dean RA, Talbot NJ, Ebbole DJ, Farman ML, Mitchell TK, Orbach MJ, Thon M, Kulkarni R, Xu JR, Pan H, Read ND, Lee YH, Carbone I, Brown D, Oh YY, Donofrio N, Jeong JS, Soanes DM, Djonovic S, Kolomiets E, Rehmeyer C, Li W, Harding M, Kim S, Lebrun MH, Bohnert H, Coughlan S, Butler J, Calvo S, Ma LJ, Nicol R, Purcell S, Nusbaum C, Galagan JE, Birren BW (2005) The genome sequence of the rice blast fungus *Magnaporthe grisea*. Nature 434:980–986

6. El Gueddari NE, Rauchhaus U, Moerschbacher BM, Deising HB (2002) Developmentally regulated conversion of surface-exposed chitin to chitosan in cell walls of plant pathogenic fungi. New Phytol 156:103–112CrossRef

7. Fujikawa T, Kuga Y, Yano S, Yoshimi A, Tachiki T, Abe K, Nishimura M (2009) Dynamics of cell wall components of *Magnaporthe grisea* during infectious structure development. Mol Microbiol 73:553–570

8. Harata K, Muraki M (2000) Crystal structures of *Urtica dioica* agglutinin and its complex with tri-*N*-acetylchitotriose. J Mol Biol 297:673–681

9. Hekmat O, Tokuyasu K, Withers SG (2003) Subsite structure of the endo-type chitin deacetylase from a deuteromycete, *Colletotrichum lindemuthianum*: an investigation using steady-state kinetic analysis and MS. Biochem J 374:369–380

10. Howard RJ, Bourett TM, Ferrari MA (1991) Infection by *Magnaporthe*: an in vitro analysis. In: Mendgen K, Lesemann DE (eds) Electron microscopy of plant pathogens. Springer, Berlin, pp 251–264

11. Kaku H, Nishizawa Y, Ishii-Minami N, Akimoto-Tomiyama C, Dohmae N, Takio K, Minami E, Shibuya N (2006) Plant cells recognize chitin fragments for defense signaling through a plasma membrane receptor. Proc Natl Acad Sci USA 103:11086–11091

12. Kankanala P, Czymmek K, Valent B (2007) Roles for rice membrane dynamics and plasmodesmata during biotrophic invasion by the blast fungus. Plant Cell 19:706–724

13. Kishimoto K, Kouzai Y, Kaku H, Shibuya N, Minami E, Nishizawa Y (2010) Perception of the chitin oligosaccharides contributes to disease resistance to blast fungus *Magnaporthe oryzae* in rice. Plant J 64:343–354

14. Li L, Ding SL, Sharon A, Orbach M, Xu JR (2007) Mirl is highly upregulated and localized to nuclei during infectious hyphal

growth in the rice blast fungus. Mol Plant Microbe Interact 20:448–458

15. Miya A, Albert P, Shinya T, Desaki Y, Ichimura K, Shirasu K, Narusaka Y, Kawakami N, Kaku H, Shibuya N (2007) CERK1, a LysM receptor kinase, is essential for chitin elicitor signaling in*Arabidopsis*. Proc Natl Acad Sci USA 104:19613–19618

16. Nishizawa Y, Nishio Z, Nakazono K, Soma M, Nakajima E, Ugaki M, Hibi T (1999) Enhanced resistance to blast (*Magnaporthe grisea*) in transgenic rice by constitutive expression of rice chitinase. Theor Appl Genet 99:383–390

17. Ramonell K, Berrocal-Lobo M, Koh S, Wan J, Edwards H, Stacey G, Somerville S (2005) Loss-of-function mutations in chitin responsive genes show increased susceptibility to the powdery mildew pathogen *Erysiphe cichoracearum*. Plant Physiol 138:1027–1036

18. Rappleye CA, Eissenberg LG, Goldman WE (2007) *Histoplasma capsulatum* α-(1,3)-glucan blocks innate immune recognition by the β-glucan receptor. Proc Natl Acad Sci USA 104:1366–1370

19. Saitoh K, Togashi K, Arie T, Teraoka T (2006) A simple method for a mini-preparation of fungal DNA. J Gen Plant Pathol 72:348–350

20. Saitoh K, Nishimura M, Kubo Y, Hayashi N, Minami E, Nishizawa Y (2008) Construction of a binary vector for knockout and expression analysis of rice blast fungus genes. Biosci Biotechnol Biochem 72:1380–1383

21. Shibuya N, Minami E (2001) Oligosaccharide signalling for defense responses in plant. Physiol Mol Plant Pathol 59:223–233

22. Tanabe S, Okada M, Jikumaru Y, Yamane H, Kaku H, Shibuya N, Minami E (2006) Induction of resistance against rice blast fungus in rice plants treated with a potent elicitor, *N*-acetylchitooligosaccharide. Biosci Biotechnol Biochem 70:1599–1605

23. van den Burg HA, Westerink N, Francoijs KJ, Roth R, Woestenenk E, Boeren S, de Wit PJGM, Joosten MHAJ, Vervoort J (2003) Natural disulfide bond-disrupted mutants of AVR4 of the tomato pathogen *Cladosporium fulvum* are sensitive to proteolysis, circumvent Cf-4-mediated resistance, but retain their chitin binding ability. J Biol Chem 278:27340–27346

24. van den Burg HA, Harrison SJ, Joosten MHAJ, Vervoort J, de Wit PJGM (2006) *Cladosporium fulvum* Avr4 protects fungal cell walls against hydrolysis by plant chitinases accumulating during infection. Mol Plant Microbe Interact 19:1420–1430

25. Wan J, Zhang XC, Neece D, Ramonell KM, Clough S, Kim SY, Stacey MG, Stacey G (2008) A LysM receptor-like kinase plays a critical role in chitin signaling and fungal resistance in*Arabidopsis*. Plant Cell 20:471–481

Chapter 11

Antifungal Activity of Selected Medicinal Plant Extract on Fusarium oxysporum Schlechtthe Causal Agent of Fusarium Wilt Disease in Tomato

Farzana Ashrafi Neela[1*], Ismat Ara Sonia[1], Shamim Shamsi[2]

[1]Department of Botany, Rajshahi University, Rajshahi, Bangladesh

[2]Department of Botany, Dhaka University, Dhaka, Bangladesh

ABSTRACT

The antifungal activity of ethanol and acetone extract of leaves of nine medicinal plants: Piper betel, Lowsonia inermis, Psidium guajava, Carica papaya, Moringa oleifera, Mimosa pudica, Catharanthus roseus, Adhatoda vasica and Andrographis paniculata against Fusarium oxysporum the causal agent of Fusarium wilt in tomato was assessed. All the extracts inhibited mycellial growth at various levels. Among them the superior inhibition (100%) was found in 15% concentration of ethanol extract of Lowsonia inermis and Psidium guajava against Fusarium. In all plant extract there were no significant differences between 20% and 25% concentration, except Piper betel, Carica papaya, Andrographis paniculata and Lawsonia inermis. Analysis of variance results on mycelial growth in different concentration shows that the item was highly significant.

Keywords: Medicinal Plants, Acetone Extract, Ethanol Extract, Inhibitory Effect, Disease Management

1. INTRODUCTION

Tomato (Lycopersicon esculentum Mill.) is economically the most important and popular vegetables throughout the world including Bangladesh. Successful cultivation of tomato is hindered by various diseases caused by plant pathogens. Fusarium wilt of tomato caused by Fusarium oxysporum f. sp. lycopersici is one of the most wide spread and destructive diseases, causing infection and losses to crop growers [1] . It is now a major concern not only in Bangladesh, but also in other regions of the world [2] .

Use of synthetic fungicides is mainly practiced for management of wilt disease [3] . This measure may cause adverse effects on the environment and human health [4] . In addition the residue of fungicides remaining in the fruit may decrease the quality of the product. The increased awareness of the environmental problems associated with fungicides has led to the search for non-conventional chemicals of biological origin for the management of this disease. Therefore, it is challenge of time to search an alternative control measure of plant diseases by using natural plant materials.

The antimicrobial activity of medicinal plant has been recognized for many years. Leaf extract of Lawsonia inermis, Achyranthes aspera, Mimosa pudica, has been reported significant antimicrobial activities [5] [6] . Plant extract may represent an ideal solution to the problem; it could be easily tested in vitro [7] . The effects of neem products have been reported to have significant controlling effects against fungal spore germination [8] . Extract of Terminalia chebula, Mangifera indica and Eucalyptus citriodora and their essential oils have been reported to be effective antimicrobials against bacteria and fungi [9] . The chloroform extract of Piper betle L. was found to be effective in controlling Fusarium spp. [10] [11] . Aqueous extracts of 46 plants against Fusarium spp. revealed that 12 plants have recorded significant antifungal activity and these plants could be exploited for eco-friendly management [10] . However, the use of plant products for control of Fusarium wilt of tomato in Bangladesh is rather limited. In this study, we reported the effectiveness of ethanol and acetone extract of some common medicinal plants in reducing populations of F. oxysporum the causal agent of Fusarium wilt in tomato in vitro.

2. MATERIALS AND METHODS

2.1. Preparation of Fungal Media (Potato Dextrose Agar)

The composition of Potato Dextrose is 4 gm/L of potato extract and 20 gm/L of dextrose. For solidification 20 gm/L of agar was added. Final pH of this media is 5.6 ± 0.2 at 25°C. These constituents were mixed, autoclaved and poured into Petri plates for solidification. These plates were used for antifungal studies.

2.2. Isolation and Identification of the Pathogen

The infected parts of tomato plant were collected in the polythene bags, which were made airtight. Collected materials were labeled properly and then brought to the Laboratory of Plant Pathology, Mycology and Microbiology, Department of Botany, University of Rajshahi, Bangladesh. The pathogen was isolated on potato dextrose agar (PDA) medium [12] . The infected parts of tomato plant were cut into small pieces. The pieces were then washed in running tap water, sterilized in 0.1 percent mercuric chloride solution and washed repeatedly for several times in sterilized distilled water to remove mercuric chloride solution. Three pieces were transferred to PDA plate. Plates were incubated at 25°C ± 2°C for 15 days for recovery of pathogen. Fusarium oxysporum was purified by single spore method and according to morphological characteristics; identification was done with the help of standard keys [13] [14] . Pathogenicity tests were done on potted tomato plants according to Koch's postulate [12] .

2.3. Preparation of Plant Materials

Fresh leaves of nine medicinal plants were collected from the surrounding areas of Motihar, Rajshahi, Bangladesh (Table 1). Plant specimens were brought to the Laboratory of Plant Pathology, Mycology and Microbiology, Department of Botany, University of Rajshahi. These plants were authenticated by Taxonomist Dr. A. H. M. Mahbubur Rahman, Associate Professor, Department of Botany, University of Rajshahi, and a voucher specimen was deposited at the Herbarium of the Department.

Leaves of plants were washed thoroughly under running tap water and soaked in 2% solution of sodium hypochlorite for 20 min, rinsed thoroughly with sterilized distilled water and air dried at room temperature. The dried plant materials were milled. 100 g of each finely ground plant materials were used for extraction in 100 ml of ethanol and acetone solvents, respectively. The solution was kept overnight at room temperature and filtered using filter papers (Whatman filter paper No. 1).

2.4. Effect of Different Concentrations of Extracts on Radial Growth of Fusarium Strain

Plant extracts which could suppress the fungal growth were tested for their efficiency against the pathogen by using an agar dilution technique. Five ml of each extract concentration was added with 95 ml of molten PDA. Thus obtained concentration of 5%, 10%, 15%, 20%

Table 1. List of plant materials used in this study.

Local name	Scientific name	Family
Pan	*Piper betel* L.	Liliaceae
Mehedi	*Lowsonia inermis* L.	Lythraceae
Payara	*Psidium guajava* L.	Myrtaceae
Pepe	*Carica papaya* L.	Caricaceae
Sojina	*Moringa oleifera* Lam.	Moringaceae
Lojjaboty	*Mimosa pudica* L.	Fabaceae
Nayontara	*Catharanthus roseus* (L.) G Don	Aacanthaceae
Basok	*Adhatoda vasica* L.	Aacanthaceae
Kalomegh	*Andrographis paniculata* (Burm.f.) Wall. Ex Nees	Aacanthaceae

and 25% extract to a PDA medium. After the solidification of the medium, 1 cm diameter of mycelia block from 7-day-old colony of F. oxysporum was inoculated in the center of each Petri plate and incubated at ±25°C. The colony diameter of F. oxysporum was measured after 7 days of incubation. Three replicates in a completely randomized design were used within each treatment. The media amended with ethanol and acetone was considered as negative and recommended fungicide sulcox was considered as positive control, respectively. The efficacy of medicinal plant products was expressed and percent of radial mycelial growth over the control which was calculated by using the following formula [7] :

Inhibition (%) = [(C – T)/C] × 100 where, C and T represent the diameter of control and treated colony, respectively. Data on mycelial growth at 3, 6, 9 and 12 days after inoculation (DAI) were recorded. To avoid bacterial contamination 0.5 g of antibacterial streptomycin was added to 1 L of PDA medium.

2.5. Statistical Analysis

The effect of ethanol and acetone extract of plants at different concentrations on the radial growth of F. oxysporum evaluated by a one way analysis of variance (ANOVA). Mean differences between treatments or concentration levels of the plant extracts were separated by Fisher's [15] test significant difference (LSD) at 5% significant probability level. The ANOVA was computation by R programming language.

3. RESULT AND DISCUSSION

For eco-friendly and sustainable management of Fusarium wilt disease, the present study tested the antifungal activity of ethanol and acetone extracts and their respective dilutions of nine selected medicinal plants based on their availability in Rajshahi and their traditional use against fungal pathogen. It was found that at 25% concentration; almost all plant extracts were effective in reducing the mycelial growth of Fusarium strain (Table 2). Effect of different plant extracts on radial mycelial growth of Fusarium sp. on 12 days of incubation (DAI) is presented in Table 3 and Table4Hundred percent inhibition at 12 DAI and 25% concentration were observed on almost all plant extract except ethanol extract of Andrographis paniculata (91%), Carica papaya (89.41%), Moringa oleifera (90.59%) and on acetone extract of Andrographis paniculata (85%), Catharanthus roseus (90.76%) and Psidium guajava (87.5%). Maximum inhibition was found at 15% concentration of ethanol extract of Lowsonia inermis and Psidium guajava against studied fungi. Representative figure was showed in Figure 1. Plant extract of Piper betel, Carica papaya, Andrographis paniculata and Lawsonia inermis did not show significant difference between 20% and 25% concentrations. Result of analysis of variance on mycelial growth in different concentration showed that the item was highly significant. LSD value also indicates significant in different concentrations of plant extract and radial growth (Table 3 and Table 4).

Table 2. Effect of different plant extracts in two solvents at 25% concentration on growth of Fusarium sp.

Plant material	% of inhibition	
	Ethanol	Acetone
P. betel	100	100
L. inermis	100	100
P. guajava	100	87.5
C. papaya	89.41	100
M. oleifera	90.59	100
C. roseus	100	90.76
M. pudica	100	100
A. vasica	100	100
A. paniculata	91	85

From the investigation, it was observed that leaf extract of Piper betel in ethanol and acetone at 20% concentrations effectively suppressed the mycelial growth of the F. oxysporum. Inhibition observed by P. betel may be due to the presence of the essential oils which contained phenolic compounds [16] [17] . There have been reports on the antifungal activities of P. betle [18] . Singha et al. [11] reported the control of fusarium wilt of tomato caused by Fusarium oxysporum using chloroform extract of P. betle leaf. They observed that it was efficient in reducing Fusarium population in soil than carbendazim. These facts suggested that betel leaf extracts exhibit significant fungicidal properties that inhibit the mycelia growth of Fusarium strain.

Table 3. Effect of various concentration of ethanol extracts on the radial mycelial growth (average mm) on Fusarium sp. at 12 DAI.

	Cconcentration of ethanol extract					
Plant material	**5%**	**10%**	**15%**	**20%**	**25%**	**LSD**
P. betel	30a*	14.76b	12.33c	0d	0d	2.101
L. inermis	6.67a	6b	0c	0c	0c	0.469
P. guajava	39.67a	9.33b	0c	0c	0c	1.558
C. papaya	48.67a	32.33b	18.33c	15.67d	8.33e	1.758
M. oleifera	34.33a	17.67b	11.67c	7.67d	7.67d	1.937
C. roseus	43.33a	14.67b	6c	0d	0d	1.558
M. pudica	37.33a	15.33b	10.33c	0d	0d	1.819
A. vasica	41a	6b	6b	0c	0c	1.409
A. paniculata	26.67a	18.67b	10c	9.67cd	8.33d	1.485

*Each treatment was replicated 3 times means followed by the same letter within a row do not differ significantly at $p > 0.05$ according to LSD test by R programming language.

Table 4. Effect of various concentration of acetone extracts on the radial mycelial growth (average mm) on Fusarium sp. at 12 DAI.

	Concentration of acetone extract					
Plant material	**5%**	**10%**	**15%**	**20%**	**25%**	**LSD**
P. betel	23.67a*	9.67b	6c	0d	0d	2.253
L. inermis	18a	6b	6b	0c	0c	0.814
P. guajava	28a	22.33b	16c	11d	8.67e	1.558
C. papaya	36.33a	20.33b	6c	0d	0d	1.329
M. oleifera	18.33a	13.33b	8.33c	6d	0e	1.993
C. roseus	42.33a	33b	29.67c	17.67d	6e	1.993
M. pudica	16.33a	8.67b	8.33b	6.33c	0d	1.243
A. vasica	35.33a	18.33b	6c	6c	0d	2.395
A. paniculata	30.33a	24.67b	17.33c	14.33c	14	3.845

*Each treatment was replicated 3 times means followed by the same letter within a row do not differ significantly at $p > 0.05$ according to LSD test by R programming language.

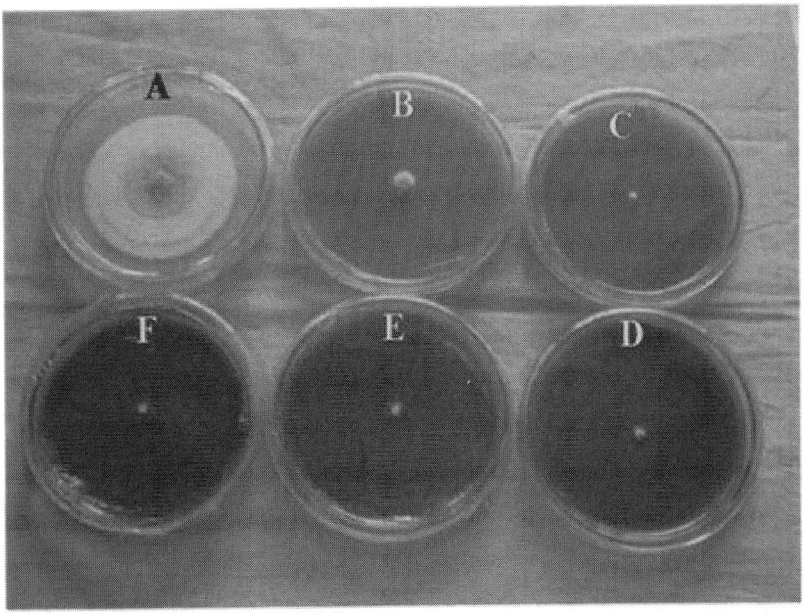

Figure 1. Superior inhibition of ethanol extract of Psidium guajava against Fusarium strain. Control (A); 5% (B); 10% (C); 15% (D); 20% (E) and 25% (F) of concentration.

The inhibitory effect of leaf of Lawsonia inermis was effective in 15% ethanol than acetone solution at 12 DAI. It is not surprising because Lawsonia inermis Linn. or henna has been used since the earliest times as a herbal medicine. Antifungal activity study with different concentrations of ethyl acetate and ethanol extract of flower and leaf of Lawsonia inermis was dose dependent and was found to be effective even at 1 mg/100µl against Penicillium notatum, Fusarium oxysporum, Colletotrichum gloeosporioides except Rhizopus stolonifer [19] . The fungicidal property of L. inermis might be due to the presence of Lawsone 2-hydroxyl-1, 4-Naphthaquino [20] .

The ethanol extract of Psidium guajava and acetone extract of Carica papaya displayed inhibitory activities, respectively. Few reports are available on these plants for developing commercial formulations for applications against fungal pathogen. Previous study demonstrated that the acetone extracts of leaves of above plants showed moderate to strong activity against Aspergillus niger, Fusarium sp. and Candida albicans [21] , whereas we have found that the ethanol extract of P. guajava showed strong inhibition at 15% concentration and acetone extract of C. papaya were at 20% concentration against Fusarium sp., respectively. The results indicated that the two plant extracts are good antifungal agents but P. guajava is more effective than C. papaya. It is therefore, encouraging to identify and characterize the active principle of these two plants.

Moringa oleifera is an exceptionally nutritious vegetable tree with a variety of potential uses [22] . The plant was reported to contain various amino acids, fatty acids, vitamins, and nutrients [22] . The seed oil has physical and chemical properties equivalent to that of olive oil and contains a large quantity of tocopherols [23] . Many reports described the antifungal and antibacterial activities of M. oleifera [24] . Ethanol extracts of M. oleifera seeds and leaves showed anti-fungal activities in vitro against dermatophytes. Furthermore, 44 chemical compounds were found in essential oil from leaves of M. oleifera [25] . The fruit extract of M. oleifera showed a broad-spectrum antifungal activity against Colletotrichum, Curvularia, Fusarium and Alternaria at different concentrations of methanol extract [22] . In this investigation, we have also found the antifungal activity of M. oleifera leaves was effective against Fusarium sp. at 25% concentration of acetone than ethanol solvent. This result is noteworthy because this extracts showed a positive relationship with the other findings.

Catharanthus roseus is well known for its pharmacological significance [26] . Antifungal potential in crude extracts of C. roseus against clinically significant fungal strains (Candida albicans, Aspergillus fumigatus, Aspergillus niger, Fusarium moniliforme) has been reported and ethanol solvent was found to be better extraction solvent as compared to acetone [26] . In this study, we have also found that ethanol extracts of leaf of C. roseus were more effective than acetone extract and reduced the mycelia growth of F. oxysporum. It showed 100% inhibition of the test fungus at 20% ethanol solution. These data depicts that the pattern of inhibition largely depend upon extraction solvent.

Ethanol and acetone leaf extract of Mimosa pudica and Adhatoda vasica at 25% concentration 100% inhibition were recorded. Methanolic extract of M. pudica exhibited least activity against Pithyum debaryanum [27] . Siva et al. [28] reported that ethanol and acetone leaves extract of Adhatoda vasica showed 100% inhibition against F. oxysporuim at 40% concentration. Present result showed that ethanol and acetone leaf extract of M. pudica and A. vasica were more effective than the observation of Ambikapathy et al. [27] and Siva et al. [28] .

Radial mycelial growth was recorded at all concentration of leaf extract of Andrographis paniculata. The lowest radial mycelial growth was measured in 25% concentration of leaf extract of A. paniculata at 12 DAI 8.33 mm in ethanol as solvent and 14 mm in acetone extract. Yasmin et al.[29] was first reported on antifungal properties of leaf extract of A. paniculata resulted 33.53% growth inhibition of Fusarium moniliforme. Present study also supported the findings of Yasmin et al. [29] that A. paniculatamay serve as candidates of plant species for their exploitation as potent fungitoxicants for controlling Fusarium wilt of tomato.

4. CONCLUSION

In conclusion, the present study laboratory screening of plant extracts has given encouraging results, indicating their potential use in the management of Fusarium wilt in tomato caused by Fusarium oxysporum. Further field trial and photochemical analysis of the active compounds of those plants would give a strong antifungal activity comparable to synthetic fungicides. The plant materials used in this study are abundant and commonly found in Rajshahi, Bangladesh.

ACKNOWLEDGEMENTS

The authors are thankful to the Department of Botany, Rajshahi University, Bangladesh for providing the necessary facilities.

REFERENCES

1. Hanaa, R.M.F., Abdou, Z.A., Salama, D.A., Ibrahim, A.R. and Sror, H.A.M. (2011) Effect of Neem and Willow Aqueous Extracts on Fusarium Wilt Disease in Tomato Seedlings: Induction of Antioxidant Defensive Enzymes. Annals of Agricultural Sciences, 56, 1-7.

2. Agbenin, N.O. and Marley, P.S. (2006) In-Vitro Assay of Some Plant Extracts against Fusarium oxysporum f. sp. lycopersici Causal Agent of Tomato. Journal of Plant Protection Research, 46, 117-121.

3. Akkopru, A. and Demir, S. (2005) Biological Control of Fusarium Wilt in Tomato Caused by Fusarium oxysporum f. sp. lycopersici by AMF Glomus intraradices and Some Rhizobacteria. Journal of Phytopathology, 153, 544-550.

4. Agbenin, N.O., Emechebe, A.M. and Marley, P.S. (2004) Evaluation of Neem Seed Powder for Fusarium Wilt and Meloidogne Control on Tomato. Archives of Phytopthology and Plant Protection, 37, 319-326.

5. Aktar, A., Neela, F.A., Khan, M.S.I., Islam, M.S. and Alam, M.F. (2010) Screening of Ethanol, Petroleum Ether and Chloroform Extracts of Medicinal Plants Lawsonia inermis L. and Mimosa pudica L. for Antibacterial Activity. Indian Journal of Pharmaceuticals Science, 72, 388-392.

6. Khan, M.S.I., Neela, F.A., Aktar, A., Rahman, M.M. and Alam, M.F (2009) Antibacterial Activity of Achyranthes aspera L.—An in Vitro Study. Journal of Environmental Science & Natural Resources, 2, 45-48.

7. Dissanayake, M.L.M.C. (2014) Inhibitory Effect of Selected Medicinal Plant Extracts on Phytopathogenic Fungus Fusarium oxysporum. Annual Research & Review in Biology, 4, 133-142.

8. Singh, U.P., Singh, H.B. and Singh, R.B. (1980) The Fungicidal Effect of Neem (A. indica) Extracts on Some Soil Born Pathogens of Gram (Cicer arietinum). Mycologia, 72, 1077-1093.

9. Parekh, J. and Chanda, S. (2006) Screening of Aqueous and Alcoholic Extracts of Some Indian Medicinal Plants for Antibacterial Activity. Indian Journal of Pharmceutical Sciences, 68, 835-838.

10. Satish, S., Raghavendra, M.P. and Raveesha, K.A. (2009) Anti-Fungal Potentiality of Some Plant Extracts against Fusarium spp. Archives of Phytopathology and Plant Protection, 42, 618-625.

11. Singha, I.M., Kalita, M.C., Wann, S.B., Kakoty, Y., Das, J., et al. (2011) Control of Fusarium wilt of Tomato Caused by Fusarium oxysporum f. sp. lycopersici Using Leaf Extract of Piper betle L.: A Preliminary Study. World Journal of Microbiology and Biotechnology, 27, 2583-2589.

12. Agrios, G. (2005) Plant Pathology. 5th Edition, Elsevier Academic Press, Amsterdam, 26-27,398-401.

13. Booth, C. (1971) The Genus Fusarium. Commonwealth Mycological Institute, Kew, Surrey, 237.

14. Nelson, P.E., Toussoun, T.A. and Marasas, W.F.O. (1983) Fusarium Species. In: An Illustrated Manual for Identification, Pennsylvania State University Press, University Park, 193.

15. Fisher, A., Immer, F.R., and Tedin, O. (1932) The Genetical Interpretation of Statistic of the Third Degree of Study of Quantitative Inheritance. Genetics, 17, 107-114.

16. Phangthip, P., Thaewnonngiw, B., Kanchanarach, W., Sangdee, A. and Rittideth, S. (2005) Antifungal Efficacy of Piper betle Linn. Extract against the Growth of Mat Fungi. 31st Congress on Science and Technology of Thailand, Bangkok, 18-20 October 2005, 3. [Paper reference:1]

17. Begum, J., Yusuf, M., Chowdhury, J.U., Khan, S. and Nural, A.M. (2007) Antifungal Activity of Forty Higher Plants against Phytopathogenic Fungi. Bangladesh Journal of Microbiology, 24, 76-78.

18. Ali, I., Khan, F.G., Suri, K.A., Gupta, B.D., Satti, N.K., Dutt, P., et al. (2010) In Vitro Antifungal Activity of Hydroxychavicol Isolated from Piper betle L. Annals of Clinical Microbiology and Antimicrobials, 9, 7-15.

19. Jeyaseelan, E.C., Vinuja, T., Pathmanathan, K. and Jeyadevan, J.P. (2012) Control of Plant Pathogenic Fungi Using Organic Solvent Extracts of Leaf, Flower and Fruit of Lowson iainermis L. International Journal of Pharmaceutical and Biological Archives, 3, 783-788.

20. Themnozhi, M. and Sivaraj, R. (2009) Phytochemical Analysis of Antimicrobial Activity of Polyalthia longifolia. International Journal of Pharma and Bio Sciences, 1, 1-7.

21. Karsha, P.V., Naseem, S. and Bhagyalakshmi, O. (2009) Antimicrobial Activity of Leaf Extracts of Guava (Psidium guajava) and Papaya (Carica papaya). Advances in Plant Sciences, 22, 429-431.

22. Sayeed, M.A., Hossain, M.S., Chowdhury, M.E.H. and Haque, M. (2012) In Vitro Antimicrobial Activity of Methanolic Extract of Moringa oleifera Lam. Fruits. Journal of Pharmacognosy and Phytochemistry, 1, 94-96.

23. Tsaknis, J., Lalas, S., Gergis, V., Dourtoglou, V. and Spilotis, V. (1999) Characterisation of Moringa oleifera Variety Mbololo Seed Oil of Kenya. Journal of Agricultural and Food Chemistry, 47, 4495-4499.

24. Gupta, R., Kannan, G.M., Sharma, M. and Flora, S.J.S. (2005) Therapeutic Effects of Moringa oleifera on Arsenic Induced Toxicity in Rats. Environmental Toxicology and Pharmacology, 20, 456-464.

25. Chuang, P.H., Lee, C.W., Chou, J.Y., Murugan, M., Shieh, B.J. and Chen, H.M. (2007) Anti-Fungal Activity of Crude Extracts and Essential Oil of Moringa oleifera Lam. Bioresource Technology, 98, 232-236.

26. Kratika, K. and Gupta, S. (2013) Antifungal Properties of Leaf Extract of Catharanthus roseus L. (G.) Don. American Journal of Phytomedicine and Clinical Therapeutics, 1, 698-705.

27. Ambikapathy, V., Gomathi, S. and Panneerselvam, A. (2011) Effect of Antifungal Activity of Some Medicinal Plants against Pythium debaryanum (Hesse). Asian Journal of Plant Science and Research, 1, 131-134.

28. Siva N., Ganesan, S. and Muthuchelian, B.N. (2008) Antifungal Effect of Leaf Extract of Some Medicinal Plants against Fusarium oxysporum Causing Wilt Disease of Solanum melogena L. Ethnobotanical Leaflets, 12, 156-163.

29. Yasmin M., Hossain, K.S. and Bashar, M.A. (2008) Effects of Some Angiospermic Plant Extracts on in Vitro Vegetative Growth of Fusarium moniliforme. Bangladesh Journal of Botany, 37, 85-88.

Chapter 12

Verticillium Tricorpus Causing Lettuce Wilt in Japan Differs Genetically from California Lettuce Isolates

Toshiyuki Usami[1] , Takeshi Kanto[2] , Patrik Inderbitzin[3] , Mizuho Itoh [1], Gan Kisaki[4] , Yoshiyuki Ebihara[5] , Wataru Suda[6] , Yoshimiki Amemiya[1] , Krishna V. Subbarao[3]

[1]*Graduate School of HorticultureChiba UniversityMatsudoChibaJapan*

[2]*Hyogo Prefectural Technology Center for Agriculture, Forestry and FisheriesKasaiJapan*

[3]*Department of Plant PathologyUniversity of California, Davis, c/o US Agricultural Research StationSalinasUSA*

[4]*Kagawa Prefectural Agricultural Experiment StationTakamatsuJapan*

[5]*Chiba Prefectural Agriculture and Forestry Research CenterSouthern Horticulture InstituteTateyamaJapan*

[6]*Graduate School of Frontier SciencesThe University of TokyoKashiwaJapan*

ABSTRACT

In January 2002, Verticillium wilt of lettuce (*Lactuca sativa* L.) caused by *Verticillium tricorpus* occurred in upland paddy fields in Hyogo Prefecture for the first time in Japan. This fungal species was first isolated from lettuce in California, USA. In the present study, the genetic relationships between the American and Japanese isolates of *V. tricorpus* from lettuce were analyzed to determine whether the pathogen could have migrated to Japan from the USA, the major lettuce-seed supplier for Japan. Nucleotide sequences of the rDNA internal transcribed spacer regions, as well as the genes coding for translation elongation factor 1-alpha and RNA polymerase II were compared among American and Japanese *V. tricorpus* isolates from lettuce. The Japanese isolates of *V. tricorpus* were distinct from the American. Random amplified polymorphic DNA analyses also supported this conclusion. These results demonstrated that Verticillium wilt on lettuce caused by *V. tricorpus* in Japan was not related to the isolates causing the disease in California.

Keywords: *Verticillium tricorpus* Verticillium wilt Lettuce Phylogenetic analysis Random amplified polymorphic DNA (RAPD)

INTRODUCTION

In January 2002, Verticillium wilt of lettuce (*Lactuca sativa* L.) occurred in upland paddy fields in Awaji Island, Hyogo Prefecture, for the first time in Japan (Kanto et al. 2003). The disease spread to neighboring Kagawa Prefecture within a few years. Phytopathogenic species of*Verticillium* cause Verticillium wilt in a variety of hosts (Pegg and Brady 2002). These species are identified primarily by the resting structures (Isaac 1967; Smith 1965) that they produce in culture; *Verticillium dahliae* Kleb., produces microsclerotia, and *V. albo-atrum* Reinke and Berthold, produces dark resting mycelia. In contrast, *V. tricorpus* I. Isaac produces microsclerotia, dark resting mycelia, and chlamydospores. *Verticillium* species isolated from diseased lettuce on Awaji Island formed all three resting structures in culture. Optimal temperature for growth of this fungus was between 20 and 22°C with the mature cultures developing yellow pigmentation. From these morphological and cultural characteristics, Kanto et al. (2005) identified the pathogen from lettuce as *V. tricorpus*.

Although Isaac (1953) described *V. tricorpus* in 1953, this species was not reported in Japan until 1998. In 1998, Ebihara et al. (2003) first isolated *V. tricorpus* from potato tubers in Japan. However, these isolates were nonpathogenic on potato, Chinese cabbage, eggplant, green pepper, larkspur, parsley, snapdragon, soybean, tobacco, and tomato. Later, Chikuo et al. (2000) found that *V. tricorpus* isolates were pathogenic on delphinium. The Verticillium wilt outbreak on lettuce in Hyogo Prefecture in 2002 caused by *V. tricorpus* is only the second incident of the disease in Japan; thus, besides being relatively new to Japan, *V. tricorpus* has also been relatively rare.

The first report of *V. tricorpus* from lettuce was from California in 1997, where it was judged to be weakly pathogenic (Qin et al. 2008). The United States is a principal lettuce-seed producing country for Japan. Contaminated lettuce seed can serve as inoculum for Verticillium wilt caused by the related species *V. dahliae* (Vallad et al. 2005). This pathogen can also be transported across long distances on seed (Atallah et al. 2010). It is thus conceivable that the 2002 lettuce wilt outbreak in Japan might have been caused by *V. tricorpus* introduced with contaminated seed from the USA. To address this question, we investigated the genetic similarity of *V. tricorpus* lettuce isolates from the USA and from Japan, using phylogenetic analyses based on three loci as well as random amplified polymorphic DNA (RAPD) fingerprinting.

MATERIALS AND METHODS

Fungal strains and culture

Table 1 presents strains of *V. tricorpus* used in this study. Each strain was maintained on potato sucrose agar [potato broth containing 2% (w/v) sucrose and 2% (w/v) agar] slants. For morphological observations using microscopy, strains were cultured on potato sucrose agar plate and incubated at 20°C for 2 or 3 weeks.

Table 1. Isolates of *Verticillium tricorpus* used for this study

Strain	Original host	Geographical location	Year isolated
MIH001	Lettuce	Hyogo, Japan	2002
MIH003	Lettuce	Hyogo, Japan	2002
Shichi6	Lettuce	Hyogo, Japan	2007
Shichi12	Lettuce	Hyogo, Japan	2007
KNty-2-2	Lettuce	Kagawa, Japan	2005
eLTS-1	Lettuce	Chiba, Japan	2000
eLTS-2	Lettuce	Chiba, Japan	2000
CE99Vt2	Lettuce	Chiba, Japan	1999
Ls.183[a]	Lettuce	California, USA	1997
Ls.432[a]	Lettuce	California, USA	2001
Ls.441[a]	Lettuce	California, USA	2001
Ls.442[a]	Lettuce	California, USA	2001
Ls.443[a]	Lettuce	California, USA	2001
NO-1	Delphinium	Kagawa, Japan	2008
MAFF 712235[b]	Delphinium	Miyagi, Japan	1999
MAFF 712236[b]	Delphinium	Miyagi, Japan	1999
CE98Vt1[c]	Potato	Japan	1998
CE20VtLe3[c]	Tomato	Chiba, Japan	2000
NBRC 31025[d,e]	Unknown	UK	Unknown

[a] Strains reported by Qin et al. (2008)

[b] Strains from National Institute of Agrobiological Sciences (NIAS) Genebank, Japan

[c] Strains reported by Ebihara et al. (2003)

[d] Strain from National Institute of Technology and Evaluation (NITE)-Biological Resource Center, Japan

[e] According to database of the NITE-Biological Resource Center, this strain was from I. Isaac who proposed *V. tricorpus* as a species

Pathogenicity test

Strain MIH001 was subcultured for 1 week in potato sucrose broth [potato broth containing 2% (w/v) sucrose] and incubated in an orbital shaker at 25°C. After filtering through cheesecloth, budding spores were collected by centrifugation for 5 min (3000×*g*). Fifty milliliters of spore suspension (5×10^6 spores/mL distilled water) was poured into each plastic pot containing a 4-week-old lettuce seedling. Lettuce cultivars Logic (Yokohama Nursery Co., Yokohama, Kanagawa, Japan), Cisco (Takii & Co., Kyoto, Japan), and Santos-2 (Fujii' Seed Co., Osaka, Japan) were used for experiments 1 and 2, experiments 3 and 4, and experiments 5 and 6, respectively. For inoculation of experiments 1 and 2, 14 and 7 seedlings, for experiments 3 and 4, 12 and 7 seedlings, and for experiments 5 and 6, 12 and 4 seedlings, respectively, were used. For each experiment, a similar number of uninoculated seedlings were maintained as controls. Inoculated and uninoculated plants were maintained in a growth chamber at 18°C and a 12-h photoperiod. The plants were watered enough everyday to maintain soil moisture. Wilting of leaves and vascular browning in cross or vertical sections of the taproot and stem were evaluated at 25, 33, 28, 33, 18, and 33 days after inoculation in experiments 1 through 6, respectively. Wilting of leaves was estimated on a scale of 0–4 where 0 = no wilting, 1 = lower leaves wilted, 2 = middle leaves wilted, 3 = upper leaves wilted, with or without yellowing, and 4 = systemic wilt or death of seedling. After surface sterilization by soaking in 70% (w/w) ethanol (for 30 s) and 0.5% (w/w) sodium hypochlorite (for 3 min), discolored stem sections were put on a water agar plate for reisolation of the pathogen from symptomatic plants. Wilt incidence was calculated as the ratio of the number of wilted plants to the total number of plants inoculated and expressed as a percentage.

Polymerase chain reaction and sequencing

Genomic DNA of each fungal strain was extracted using a standard phenol–chloroform extraction method (Usami et al. 2002, 2005) or a QIAamp Tissue Kit (Qiagen GmbH, Hilden, Germany). The genomic region of ribosomal DNA ITS1-5.8S-ITS2 was amplified using PCR primers ITS1 and ITS4 (White et al. 1990). A part of the translation elongation factor 1-alpha gene (*EF1α*) was amplified using primers EF1-983F and EF1-1567R (Rehner and Buckley 2005). The DNA-dependent RNA polymerase II largest subunit gene (*rpb1*) was partially amplified using PCR primers Vrpb1F and Vrpb1R, which we designed from the sequence of *V. tricorpus*(AY555904) reported by Pantou et al. (2005). Nucleotide sequences of primers are presented in Table S1 in Electronic Supplementary Material. For amplification of each genomic region, a 20-μL PCR mixture containing 20 ng fungal genomic DNA, 100 pmol of each primer, and 10 μL of*Premix Ex Taq* Hot Start Version (Takara Bio Inc., Otsu, Shiga, Japan) was prepared. The following PCR program was used for amplification of the ITS1-5.8S-ITS2 region: 95°C for 5 min; 40 cycles of 95°C for 1 min, 53°C for 1 min, and 72°C for 1 min; and 72°C for 5 min. The following PCR program was used for amplification of *EF1α* and *rpb1*: 95°C for 5 min; 40 cycles of 95°C for 1 min, 60°C for 1 min, and 72°C for 1 min; and 72°C for 5 min. The nucleotide sequence of each PCR product was analyzed using BigDye Terminator v3.1 Cycle Sequencing Kit (Life Technologies, Carlsbad, CA, USA) and an auto-sequencer (Applied Biosystems 3730 DNA Analyzer, Life Technologies).

Phylogenetic analysis

In addition to the nucleotide sequences that were determined in this study, sequences of the ITS1-5.8S-ITS2 region of *V. tricorpus* strains Ls.183 (GQ336717), Ls. 443 (GQ336792), CE98Vt1 (AB353343), CE20VtLe3 (AB353344), and *V. dahliae* TV103 (AB353345) were downloaded from the GenBank/EMBL/DDBJ database (http://www.ddbj.nig.ac.jp/index-e.html) and included in the phylogenetic analyses. Combined sequences of ITS1-5.8S-ITS2,*EF1α*, and *rpb1* regions were aligned using the CLUSTAL W program (http://clustalw.ddbj.nig.ac.jp/top-e.html) (Thompson et al. 1994) with default parameters. A maximum parsimony (MP) tree was constructed with the close-neighbor interchange method (search level = 3, random addition with 10 replications) and treating gaps as missing data, using the software MEGA 4 (http://www.megasoftware.net/) (Tamura et al. 2007). A neighbor-joining (NJ) tree was constructed using maximum composite likelihood. Bootstrap analyses for 1000 replicates were performed on each tree.

RAPD analysis

For RAPD analysis, 16 primers included in the DNA Oligomer-12 sets (Wako Pure Chemical Industries, Osaka, Japan) were used. Nucleotide sequences of the primers are presented in Table S1 of the Electronic Supplementary Material. For amplification, a 20-µL PCR mixture containing 20 ng fungal genomic DNA, 100 pmol of primer, and 10 µL of *Premix Ex Taq* Hot Start Version (Takara Bio) was prepared. The following PCR program was used for RAPD analyses: 95°C for 5 min; 40 cycles of 95°C for 1 min, 40°C for 1 min, and 72°C for 2 min; and 72°C for 5 min. Amplified products were separated on a 2% (w/v) agarose gel containing 1× Tris–acetate–EDTA (TAE), then stained with ethidium bromide and observed under UV irradiation.

RESULTS

Characteristics of fungi isolated from diseased lettuce

Verticillium wilt caused severe browning and wilt of mature lettuce in upland paddy fields on Awaji Island, Hyogo Prefecture (Fig. 1a). Vascular browning developed at the crown (Fig. 1b) and often at the stem above the crown of the diseased lettuce plants. The fungus isolated from diseased lettuce was identified as *V. tricorpus* based on its morphological and cultural characteristics (Kanto et al. 2003, 2005) including verticillate phialides (Fig. 1d) and the presence of

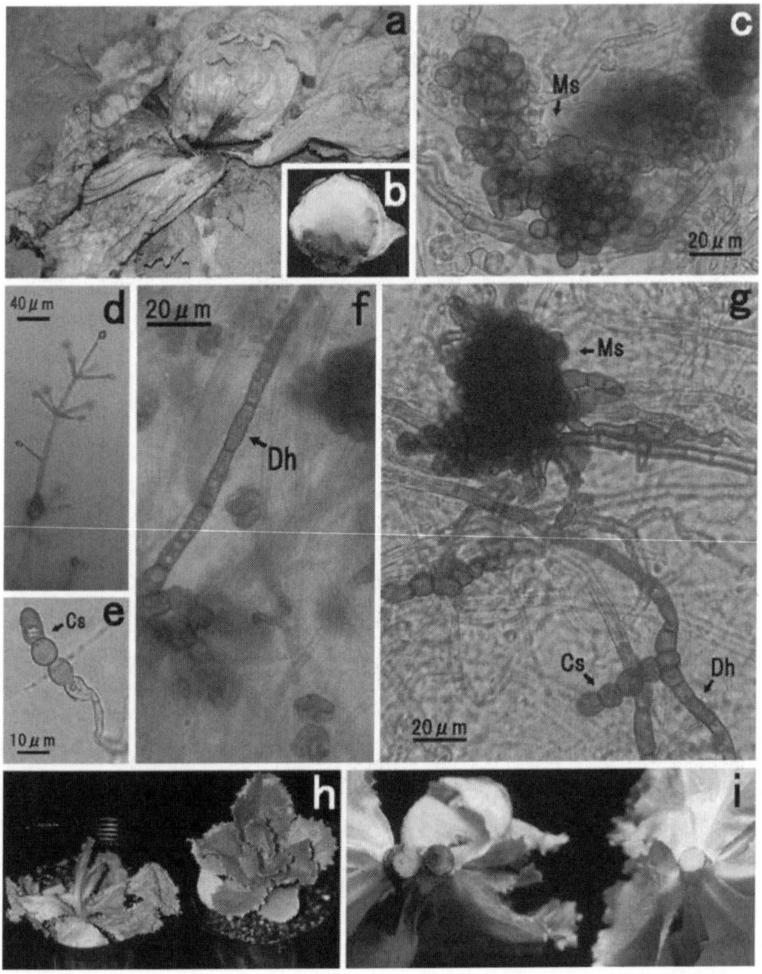

Figure 1. **a** Verticillium wilt on lettuce in an upland paddy field in Awaji Island, Hyogo Prefecture. **b** Vascular browning at the crown of the diseased lettuce. **c** Microsclerotia (Ms), **d** conidiophore, **e**chlamydospores (Cs), and **f** dark resting hyphae (Dh) of strain MIH001. **g** Microsclerotium (Ms), chlamydospores (Cs), and dark resting hyphae (Dh) of strain MIH003. **h** Seedlings of lettuce cv. Santos-2 inoculated with strain MIH001 (*left*, wilting score = 3) and noninoculated control (*right*) in a pathogenicity test (experiment 5). **i** Cross sections at crown of lettuce cv. Santos-2 inoculated with strain MIH001 (*left*) and noninoculated control (*right*) in a pathogenicity test (experiment 5)

microsclerotia (Fig. 1c, g), dark resting hyphae (Fig. 1f, g), and chlamydospores (Fig. 1e, g). However, the three types of resting structures were not always formed simultaneously. Often, only one or two kinds of resting structures were observed in one culture. Mature cultures often developed yellow pigmentation. The morphology of strain MIH001 was examined in detail as a representative of the Japanese lettuce isolates. The following mean dimensions were obtained, each based on 50 measurements: conidia 6.2 × 3.3 μm; dark resting hyphae, 5.8 μm (width); microsclerotia, 91.2 × 46.7 μm; chlamydospores, 7.7 × 6.4 μm.

Pathogenicity tests

During six pathogenicity tests using the Japanese lettuce isolate MIH001 (Table 2), typical wilting and vascular browning at the crown developed (Fig. 1h, i). Although wilting scores of 1–3 were recorded for several seedlings inoculated with strain MIH001, no seedlings had a wilting score of 4. *Verticillium tricorpus* was reisolated from symptomatic lettuce seedlings.

DNA sequencing analyses

Nucleotide sequences determined in this study were registered with GenBank/EMBL/DDBJ database (http://www.ddbj.nig.ac.jp/index-e.html). Sequences of the ITS1-5.8S-ITS2 region,*EF1α* and *rpb1* of each strain are available as accessions AB545875–AB545895, AB545897–AB545906, AB545911–AB545921, AB545923, and AB548747–AB548758.

Table 2. Results of experiments (Exp.) to test the pathogenicity of *Verticillium tricorpus* MIH001 on lettuce cultivars

Cultivar	Exp.	Inoculation	Wilted plants (%)[a]	Average wilt severity	Vascular browning (%)[b]
Logic	1[c]	MIH001	78.6	1.4	85.7
		None	0.0	0.0	0.0
	2[d]	MIH001	100.0	1.6	100.0
		None	0.0	0.0	0.0
Cisco	3[e]	MIH001	50.0	0.8	66.7
		None	0.0	0.0	0.0
	4[d]	MIH001	100.0	1.4	57.1
		None	0.0	0.0	0.0
Santos-2	5[f]	MIH001	83.3	1.2	66.7
		None	0.0	0.0	0.0
	6[g]	MIH001	100.0	1.5	100.0
		None	0.0	0.0	0.0

[a] Percentage of wilted plants
[b] Percentage of plants with vascular browning in cross and vertical sections of taproot and stem
[c] Fourteen seedlings were used for each treatment. Symptoms were evaluated at 25 d after inoculation (dai)
[d] Seven seedlings were used for each treatment. Symptoms were evaluated at 33 dai
[e] Twelve seedlings were used for each treatment. Symptoms were evaluated at 28 dai
[f] Twelve seedlings were used for each treatment. Symptoms were evaluated at 18 dai
[g] Four seedlings were used for each treatment. Symptoms were evaluated at 33 dai

Phylogenetic analyses

The maximum parsimony (MP) tree derived from combined nucleotide sequences of the ITS1-5.8S-ITS2 region, *EF1α*, and *rpb1* is presented in Fig. 2. Although two different methods [MP and neighbor-joining (NJ)] were used to derive the trees, the topologies of the trees were identical. The nucleotide sequences that we analyzed were identical among the eight Japanese lettuce isolates, except for two nucleotides in *rpb1*. No difference was found among the nucleotide sequences of American lettuce isolates and strain NBRC 31025. In contrast, the sequences differed significantly between Japanese and American lettuce isolates in all three genetic regions. Strains MIH001 and Ls.183 differed in 4 nucleotides (nt) for ITS1-5.8S-ITS2 (433 bp), 7 nt for *EF1α* (436 bp), and 9 nt for *rpb1* (632 bp). Nucleotide sequences of the ITS1-5.8S-ITS2 region and *EF1α* were identical among Japanese delphinium isolates and indistinguishable from those of American lettuce isolates. However, the *rpb1* sequence differed between the Japanese delphinium isolates and the American lettuce isolates. In comparison with the American lettuce isolates, Japanese delphinium isolates from Miyagi Prefecture (MAFF 712235 and MAFF 712236) and an isolate from Kagawa Prefecture (NO-1) differed in 7 and 2 nt. Consequently, of the three genomic regions analyzed, *rpb1* appeared to have differed the most. Japanese isolates from solanaceous plants (CE98Vt1 from potato and CE20VtLe3 from tomato) were distinguishable from other isolates by the sequence of the ITS1-5.8S-ITS2 region and *rpb1*. Their *EF1α* sequences were identical to that of the American lettuce isolates.

Figure 2. A maximum parsimony tree derived from the combined nucleotide sequences of ITS1-5.8S-ITS2 region, *EF1α*, and *rpb1*. Bootstrap values (1,000 replicates) are indicated on branches. *Vertical bars*indicate Japanese and American lettuce isolates, respectively. All other ingroup isolates are*Verticillium tricorpus* from non-lettuce hosts. *V. dahliae* strain TV103 was used as the outgroup

RAPD analyses

As examples, the RAPD patterns with primers E10, G21, C52 and D62 are shown in Fig. 3. With all 16 primers used, banding patterns of Japanese lettuce isolates were clearly similar to each other and specific to these isolates. Banding patterns of American lettuce isolates were also mutually similar. Although the non-lettuce strains NBRC 31025, MAFF 712235, MAFF 712236, and NO-1 tended to have patterns similar to that of American lettuce isolates, some differences were observed. In the Japanese isolates obtained from delphinium, banding patterns differed between isolates from Miyagi Prefecture (MAFF 712235 and MAFF 712236) and an isolate from Kagawa Prefecture (NO-1). In addition, patterns for Japanese isolates from solanaceous plants (CE98Vt1 from potato and CE20VtLe3 from tomato) resembled each other, differing from all other banding patterns. The similarities in these banding patterns correlated well with the groups obtained in the phylogenetic tree (Fig. 2). These results demonstrated that, regardless of the primer pair used, banding patterns always differed between the Japanese and American lettuce isolates.

DISCUSSION

Isolates of *V. tricorpus* capable of infecting lettuce were first reported in the USA (Qin et al.2008) followed by Japan (Kanto et al. 2003, 2005). The results of this study clearly demonstrate that the isolates of *V. tricorpus* from lettuce in Japan differed genetically from those collected in the USA. The Verticillium wilt on lettuce caused by *V. tricorpus* in Japan was therefore unrelated to the isolates causing the disease in California.

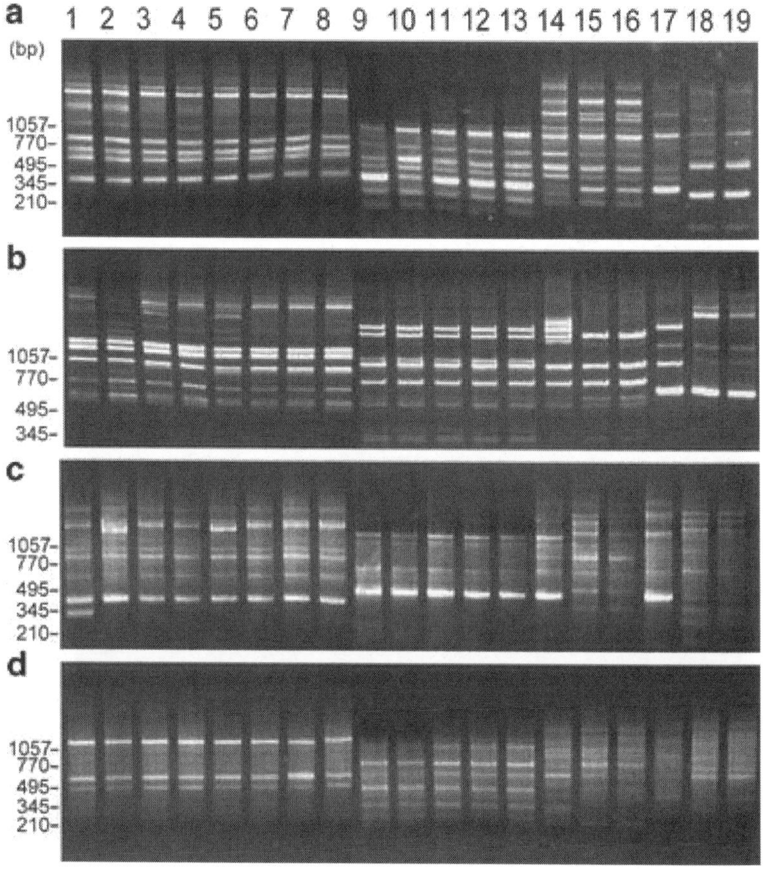

Figure 3. Random amplified polymorphic DNA (RAPD) analysis of *Verticillium tricorpus* using four primers: **a** primer E10, **b** primer G21, **c** primer C52, **d** primer D62. *Lanes 1–8* are Japanese lettuce isolates (MIH001, MIH003, Shichi6, Shichi12, KNty-2-2, CE99Vt2, eLTS1, and eLTS2). *Lanes 9–13* are American lettuce isolates (Ls.183, Ls.432, Ls.441, Ls.442, and Ls.443). *Lanes 14–19* show data for non-lettuce isolates (NBRC 31025, MAFF 712235, MAFF 712236, NO-1, CE98Vt1, and CE20VtLe3)

The genus *Verticillium* contains several important vascular wilt pathogens, but *V. tricorpus* is generally not considered one of them. In California, *V. tricorpus* was reported as a weak pathogen on lettuce

(Qin et al. 2008). However, *V. tricorpus* strain MIH001, obtained from diseased lettuce in Japan, was quite severe on some cultivars of lettuce (Table 1). Other Japanese isolates, except for three strains from Chiba Prefecture (eLTS1, eLTS2, and CE99Vt2) were as severe as MIH001 on lettuce cultivars (Kanto et al. 2003, 2005; data not shown). Seedlings infected by Japanese lettuce isolates developed both vascular browning and foliar wilting. The relatively severe symptoms caused by the Japanese *V. tricorpus* isolates in the field might be due in part to the particular environmental conditions. As opposed to the dry growing conditions prevalent in California, lettuce is often planted in very moist environments such as the upland paddy fields in Hyogo Prefecture in Japan. We are therefore unable to speculate about any potential differences in virulence between American and Japanese lettuce isolates without additional pathogenicity tests using isolates from both countries under identical conditions. Such studies may also offer clues to the conditions that may cause a generally weak pathogen on most hosts (Qin et al. 2008) to be came a pathogen of serious agricultural importance.

The differences in virulence between the Japanese and California isolates of *V. tricorpus* could also be due to genetic differences. Even though the isolates of both groups had morphologically similar resting structures, like those of the California isolates of *V. tricorpus* described by Qin et al. (2008), the nucleotide sequences of the three sequenced regions differed significantly, as did their RAPD profiles. Phylogenetic analyses of the combined data from the ITS1-5.8S-ITS2 region and *EF1α* and *rpb1* showed that Japanese and American lettuce isolates clustered into two separate clades. The eight Japanese isolates differed by only two substitutions and received maximal bootstrap support. The California isolates and an isolate from an unknown host from the UK were identical at the three loci sequenced, differed from a delphinium isolate from Japan by only two substitutions, and were part of a well-supported clade together with two additional Japanese delphinium isolates. Two *V. tricorpus* isolates included from potato and tomato isolated in Japan, formed a separate group differing from all other *V. tricorpus* isolates by at least 20 substitutions. The results from phylogenetic analyses were consistent with the RAPD data. All the RAPD banding patterns differed greatly between American and Japanese lettuce isolates, confirming that Japanese isolates are genetically distinguishable from American isolates.

These data clearly showed that the Japanese and California lettuce isolates included here represent two genetically different groups. We can thus rule out contaminated seed from California as the source for the Verticillium wilt outbreak of lettuce investigated here. Another important finding of this study is the high diversity within *V. tricorpus*, in agreement with Korolev and Katan (1999), who found considerably more vegetative compatibility groups in *V. tricorpus* than in other *Verticillium* species. Future studies will focus on comparing the virulence of the Japanese and the California isolates, as well as on further exploring the genetic diversity within *V. tricorpus.*

REFERENCES

1. Atallah ZK, Maruthachalam K, du Toit L, Koike ST, Davis RM, Klosterman SJ, Hayes RJ, Subbarao KV (2010) Population analyses of the vascular plant pathogen *Verticillium dahliae*detect recombination and transcontinental gene flow. Fungal Genet Biol 47:416–422

2. Chikuo Y, Hagiwara H, Kimijima E, Kikuchi O, Takahashi T, Izutsu S (2000) Verticillium wilt of*Delphinium* sp. caused by *Verticillium tricorpus* (Abstract in Japanese). Jpn J Phytopathol 66:303

3. Ebihara Y, Nagao H, Uematsu S, Moriwaki J, Kimishima E (2003) First report of *Verticillium tricorpus* isolated from potato tubers in Japan. Mycoscience 44:481–488

4. Isaac I (1953) A further comparative study of pathogenic isolates of *Verticillium*: *V. nubilum*Pethybr. and *V. tricorpus* sp. nov. Trans Br Mycol Soc 36:180–195

5. Isaac I (1967) Speciation in *Verticillium*. Annu Rev Phytopathol 5:201–222

6. Kanto T, Aino M, Maekawa K, Tanaka H (2003) The first occurrence of Verticillium wilt on lettuce in Japan (Abstract in Japanese). Jpn J Phytopathol 69:63

7. Kanto T, Usami T, Ebihara Y, Maekawa K, Aino M (2005) Identification of pathogens of lettuce Verticillium wilt (Abstract in Japanese). Jpn J Phytopathol 71:214–215

8. Korolev N, Katan T (1999) Vegetative compatibility grouping in *Verticillium nigrescens* and *V. tricorpus*. Mycol Res 103:65–76

9. Pantou MP, Strunnikova OK, Shakhnazarova VY, Vishnevskaya NA, Papalouka VG, Typas MA (2005) Molecular and immunochemical phylogeny of *Verticillium* species. Mycol Res 109:889–902

10. Pegg GF, Brady BL (2002) Verticillium wilts. CAB International, Oxford

11. Qin QM, Vallad GE, Subbarao KV (2008) Characterization of *Verticillium dahliae* and *V. tricorpus* isolates from lettuce and artichoke. Plant Dis 92:69–77

12. Rehner SA, Buckley E (2005) A *Beauveria* phylogeny inferred from nuclear ITS and EF1-α sequences: evidence for cryptic diversification and links to *Cordyceps* teleomorphs. Mycologia 97:84–98

13. Smith HC (1965) The morphology of *Verticillium albo-atrum*, *V. dahliae* and *V. tricorpus*. N Z J Agric Res 8:450–478

14. Tamura K, Dudley J, Nei M, Kumar S (2007) MEGA4: Molecular Evolutionary Genetics Analysis (MEGA) software version 4.0. Mol Biol Evol 24:1596–1599

15. Thompson JD, Higgins DG, Gibson TJ (1994) CLUSTAL W: improving the sensitivity of progressive multiple sequence alignment through sequence weighting, position-specific gap penalties and weight matrix choice. Nucleic Acids Res 22:4673–4680

16. Usami T, Abiko M, Shishido M, Amemiya Y (2002) Specific detection of tomato pathotype of *Verticillium dahliae* by PCR assays. J Gen Plant Pathol 68:134–140

17. Usami T, Shishido M, Ebihara Y, Kamigahira Y, Amemiya Y (2005) Retrotransposon-like elements in the genome of *Verticillium dahliae* may be used as DNA markers for fungal species and pathotypes. J Gen Plant Pathol 71:117–123

18. Vallad GE, Bhat RG, Koike ST, Ryder EJ, Subbarao KV (2005) Weedborne reservoirs and seed transmission of *Verticillium dahliae* in lettuce. Plant Dis 89:317–324

19. White TJ, Bruns T, Lee S, Taylor J (1990) Amplification and direct sequencing of fungal ribosomal RNA genes for phylogenetics. In: Innis MA, Gelfand DH, Sninsky JJ, White TJ (eds) PCR protocols: a guide to methods and applications. Academic Press, San Diego, pp 315–322

Chapter 13

Farmer's Perception of Cassava Mosaic Disease, Preferences and Constraints in Lupaula Province of Zambia

P. C. Chikoti[1*], R. Melis[2], P. Shanahan[2]

[1]*Zambia Agriculture Research Institute, Mount Makulu Research Station, Chilanga, Zambia*

[2]*University of KwaZulu-Natal, African Center for Crop improvement, Scottsville, South Africa*

ABSTRACT

Cassava is the principal staple root crop, providing a major source of calories for rural and urban households in Luapula province of Zambia. However, the yields on smallholder farms are relatively low largely due to disease infections. The study was therefore conducted to establish farmers' perceptions, knowledge and management of cassava mosaic disease (CMD), which is one of the major diseases of cassava, and to establish farmers' preferred traits, constraints and assess sources of cassava cuttings for planting. Focus group discussions (FGD) and structured interviews involving 156 farmers in Mwense, Mansa and Samfya districts were conducted from December 2008 to March 2009. Knowledge of CMD was limited among the respondents. Only 2.4% of the respondents were aware of the disease despite high CMD incidence in farmers' fields. The majority of the farmers were aware of the importance of insect pests; however, they could not differentiate between damages due to diseases or insect pests. High yield and early bulking traits were highly ranked. Most of the farmers planted local landraces on small-fields (<1 ha). It was evident that a local breeding programme aimed at developing locally adapted disease and pest resistant cassava cultivars was a pressing requirement.

Keywords: Cassava Mosaic Disease, Farmer, Knowledge

1. INTRODUCTION

Cassava is one of the major staple and most highly valued root crops in Zambia. It is mostly grown in Northern, Luapula, North-Western and Western provinces (often referred to as cassava belt) which accounts for 95% of total production [1] . Thirty percent of the population in Zambia is directly or indirectly dependent on cassava for their livelihood, with the majority from the cassava belt region [2] . In the last few years, cassava promotion and production has spread to other parts of the country such as Central, Eastern and Southern provinces.

Cassava mosaic disease has been reported to be one of the most limiting constraints to cassava production in Africa [3] . The disease in Zambia is prevalent in most of the farmers' fields [4] , affecting both local and improved cultivars. Although improved cassava cultivars have been developed by the Zambia Agriculture Research Institute (ZARI) and have been promoted by public and private extension services, including non-governmental organisations (NGOs), adoption has been slow and consequently overall production is still low, with an average yield of 5.8 $t \cdot ha^{-1}$ [5] . The adoption of a cultivar depends on the presence of farmer preferred traits among the available cultivars and the availability of information upon which decisions are based.

To assess the usefulness of any given cultivar, there is a need to determine the attributes and constraints that are responsible for farmers' choices through participatory approaches. Farmers have local knowledge on the attributes of their cultivars. Studies by Agwu and Anyaeche [6] indicated that a farmer's decision to use a particular cassava cultivar was influenced by a number of factors, some of which are trait based (high yield, low cyanide, early maturity, and colour of roots).

Improving cassava production among smallholder producers through the introduction of improved cultivars requires farmer involvement in the early stages of breeding. In many national breeding programmes where the farmers have been involved in the breeding process, improvements have been observed in the adoption and release of new cultivars. For instance, in Ghana, scientists working in collaboration with farmers identified 129 superior accessions from a total of 1350 seedlings [7] . The participatory approach improves the adoption rates through integrating local knowledge into research through dialogue between farmers and scientists. Furthermore, it is necessary for evaluating traits most preferred by the farmers. Therefore, a participatory rural appraisal (PRA) was conducted to gather information on farmers' preferences, perception, and knowledge of CMD and other production constraints and to lay the foundation for the development of CMD resistant cultivars in Zambia. The objectives of this study were: 1) to assess farmers' knowledge and perceptions of CMD; 2) to evaluate farmers' knowledge on the management of CMD; 3) to establish farmers' preferred traits and various constraints to cassava production; and 4) to assess sources of cassava cuttings for planting.

2. METHOD

2.1. Description of Study Area

The study was carried out in Mansa, Mwense and Samfya districts of Luapula province, located between latitude 8° to 12° south of the equator and longitude 28° to 30° east of Greenwich Mean Time [8] . The districts within the province are located in the high rainfall agroecological zone (AEZ III) (Figure 1) and receive above 1000 mm of rainfall per year. The rainfall pattern is monomodal and lasts from November to April. The mean annual minimum temperature is 10°C and the mean annual maximum is 31°C. The length of growing season is approximately 160 to 170 days (November to April) for rain grown crops.

2.2. Selection of Participants

The study was conducted between December 2008 and March 2009 in collaboration with the Department of Agriculture which falls under the Ministry of Agriculture and Cooperatives (MACO). A list of cassava farmers was drawn up and 20 to 25 farmers were randomly selected for the focus group discussion (FGD). In each district, 6 to 8 villages were targeted for the FGD. Adult men (56.1%) and women (43.9%) farmers were involved in the study (Table 1).

In the second stage, 90 randomly selected farmers (30 per district) were involved in the semi-structured interview (Table 2). The farmers were asked similar questions as in the FGD on their perceptions of pests and diseases, production and marketing constraints, control strategies and cropping system using a semi-structured questionnaire.

2.3. Data Collection

Prior to data collection a multi-disciplinary team was constituted comprising the principal investigator, three assistants, an extension officer (from District Agriculture Coordinators' office), and a camp officer from each study area. All the team members underwent training on administering questionnaires and handling of FGD. In addition, the participatory appraisal team reviewed interviewing techniques and questions in the questionnaire. Furthermore, the team

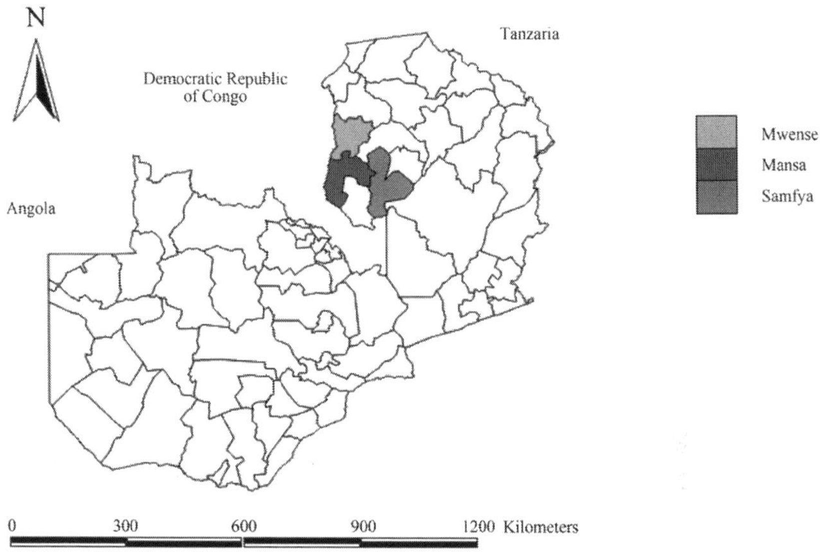

Figure 1. Districts in Zambia surveyed for the participatory rural appraisal study.

Table 1. Number of farmers by gender participating in focus group discussions held in three districts of Luapala province, Zambia.

District	Number of villages	Male	Female	Total
Mwense	6	12	10	22
Mansa	6	17	7	24
Samfya	8	8	12	20
Total	20	37 (56.1%)	29 (43.9%)	66

Table 2. Total number of farmers by gender participating in the structured questionnaire in three districts of Luapala province, Zambia.

District	Male	Female	Total
Mwense	21	9	30
Mansa	23	7	30
Samfya	23	7	30
Total	67 (74.4%)	23 (25.6%)	90

also discussed various options to use in order to extract maximum information from the farmers. The open-ended questions that were formulated allowed farmers to give their opinions freely. The interviews and discussions were conducted in the local language (Bemba) as most of the farmers were conversant in it and this encouraged wide participation.

To assess farmers' reaction and knowledge of diseases and insect pests they were presented with plants having CMD symptoms or infested with insect pests. In addition, probing and iterative techniques were used during FGD discussion and structured interviews. Some of the questions asked were repeated and rephrased to enable farmers to better understand and respond as fully as possible. Repeating and rephrasing of questions is often necessary when the study group comprises semi-literate respondents. Other techniques used were listing and ranking of constraints, observations of cultivars in the field, listing of traits, and ranking.

Data on farmers' knowledge and perception of insect pests and diseases were collected from the FGD and structured interviews. The questions centred on farmers' awareness of insect pests and diseases, cultivars grown, production and marketing constraints, cropping system used and cropping calendar. Farmers were asked to list cultivars and provide their attributes. Comparisons between the factors considered e.g. constraints, insects and diseases, were done using the ranking method. The factor with the highest number of points was ranked as first and that with lowest points last. In Mwense and Samfya districts, farmer training centres (FTCs) were used for the FGD as the study was done during the rainy season.

2.4. Data Analysis

Statistical analysis for quantitative data was analysed using the statistical package for social sciences (SPSS) [9] . Descriptive statistics, analysis of variance and mean comparisons for each district were generated.

3. RESULTS

3.1. Land Size

The average area planted with cassava by most of the farmers (46.3%) in the three districts was less than 1 ha. The rest of the cassava fields were between 1.0 to1.5 ha (22.5%) and 1.5 to 2.0 ha (31.2%). In Samfya district (Figure 2) 38.8% of the respondents had cassava fields of less than 1 ha. In Mansa, 44.4% of the respondents had fields measuring between 1.0 - 1.5 ha.

3.2. Knowledge and Perception of Cassava Mosaic Disease

In all the three districts, the majority of the respondents (97.6%) were not familiar with the symptoms of CMD and could therefore not identify the disease. Only a few respondents (2.4%) were familiar with CMD and its symptoms by virtue of working at, and being within the vicinity of, Mansa Research station. A number of reasons were given by the respondents as to the probable cause of CMD. The majority of the respondents (73.6%) thought the symptoms were as a result of harvesting of cassava leaves. On the other hand, 12.6% were of the view that CMD was caused by mealybug infestation. Other respondents were of the view that CMD was caused by old age of the plants (4.6%), cold (3.4%), and lack of rain (3.4%). The rest of the respondents attributed the cause of CMD to lack of hygiene (non-removal of affected plants). In Samfya district some farmers were able to differentiate between symptoms of mealybug (Phenacoccus manihoti Matile-Ferrero) infestation and CMD. However, the farmers did not have a name for the condition in plants that exhibited CMD symptoms.

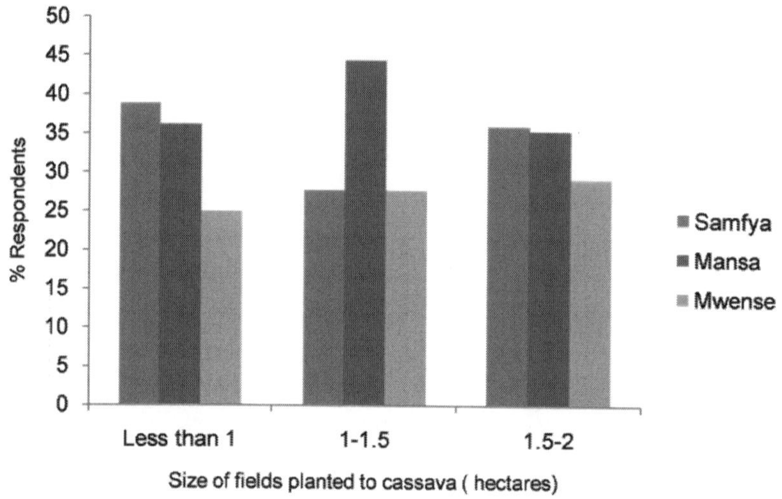

Figure 2. Size of cassava fields in Samfya, Mansa and Mwense districts.

3.3. Incidence and Severity of Cassava Mosaic Disease in Farmers' Fields

Although the farmers were not aware of CMD, symptoms of the disease were present in most of the fields. There were no significant differences in CMD incidence in Samfya, Mansa and Mwense districts. The average incidence of CMD across the three districts was 61.2%. Samfya (68.6%) and Mwense (62.6%) districts had high CMD incidence. Mansa had the least incidence (57.8%). The CMD severity was moderate in Samfya (2.4), Mansa (2.5) and Mwense (2.5) with an overall mean of 2.5.

3.4. Insect Pests of Cassava

All the fields visited had mealybug, whitefly (Bemisia tabaci Gennadius, a vector of CMD) and cassava green mite (Mononychellus tanajoa Bondar). In all three districts farmers revealed that no cultivar was resistant to these pests and that all the cassava cultivars were equally affected. When asked about pests in general, the farmers could not differentiate between damage due to diseases versus insects. The majority (91.1%) of the respondents recognised insect pests and diseases as important in their cassava fields. On the other hand, a few farmers (8.9%) thought that insect pests and diseases were not that important. Across the three districts, cassava mealybug was regarded as the most important pest by 71.0% of the farmers, whereas 15.0% of the respondents considered termites (observed by the farmers to occur during the dry season (April-October)) to be the least important. Though termites were not considered to be the major pest, they were regarded as important by 57.1% and 35.7% of the respondents in Samfya and Mansa districts, respectively. However, 7.1% of respondents in Mwense district regarded termites to be least important. Other pests that farmers mentioned were moles (Fukomys anselli) (6.5%), grasshoppers (4.3%) and cutworms (3.2%). Mealybugs were also mentioned as important pests in Mansa (32.3%), Mwense (30.8%) and Samfya districts (36.9%).

3.5. Sources of Cassava Cuttings for Planting

In Mansa district 40.0% of the farmers obtained cuttings for planting from MACO, while 36.1% of the respondents from the same district accessed cuttings for planting from their own fields. In Mwense district 41.5% of the farmers accessed cuttings for planting from fellow farmers and 26.7% of the farmers obtained cuttings for planting from MACO. In Samfya district, 39.0% of the farmers obtained cuttings for planting from their colleagues, while the rest obtained cuttings for planting from MACO (33.3%) and own fields (34.7%) [10] [11] .

3.6. Management of Cassava Mosaic Disease

From the FGD and structured interview it was evident that not a single farmer had a control or management strategy for CMD. Since the disease was poorly understood by most of the farmers, corresponding management options were not mentioned. Some farmers practised field sanitation through the removal of affected leaves, although this was intended for the control of cassava mealybug.

3.7. Cassava Cultivars Grown

The cultivars were mostly (>90%) local landraces. Across the three districts, 58.1% of the respondents grew local cultivars, 19.8% grew improved ones, while 22.1% grew both local and improved cultivars. On average four to five different cultivars were grown on each farm. The most popular cultivars included Bangweulu (20.1%), Katobamputa (19.1%), and Kabala (11.1%) (Figure 3). Farmers indicated that the improved cultivars (Mweru, Chila, Kapumpa, Kampolombo and Tanganyika) were not readily available in their localities.

3.8. Farmers' Preferred Characteristics

Across the three districts, 37% of the respondents preferred cultivars that were high yielding, while 36% preferred early bulking cultivars (Figure 4) (early bulking according to the farmers was cultivars that were ready for harvesting between 2 to 3 years after planting). Few respondents (13.4%) based their preference on the colour of cassava flour. Some respondents preferred local cultivars to improved ones as they produce better flour in terms of colour. A minority (2.4%) grew specific cultivars because the plants gave them more cuttings for planting and leaves for relish. Few respondents (1.2%) preferred cultivars with insect pest resistance.

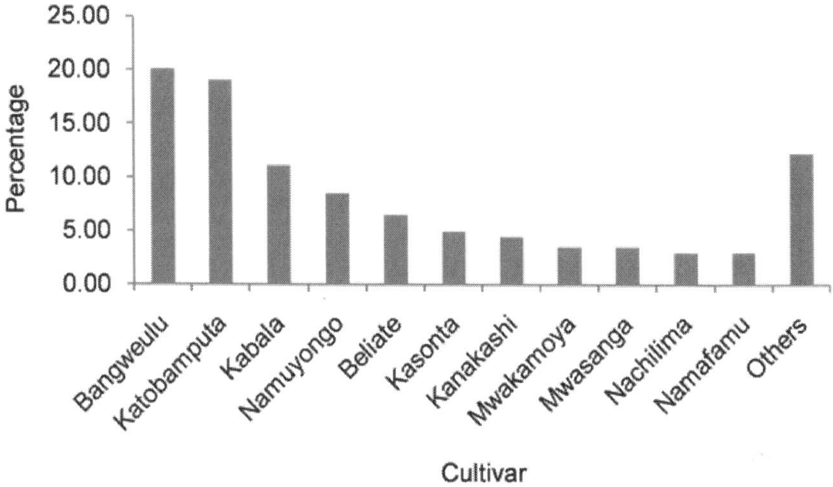

Figure 3. Cassava cultivars grown by the farmers.

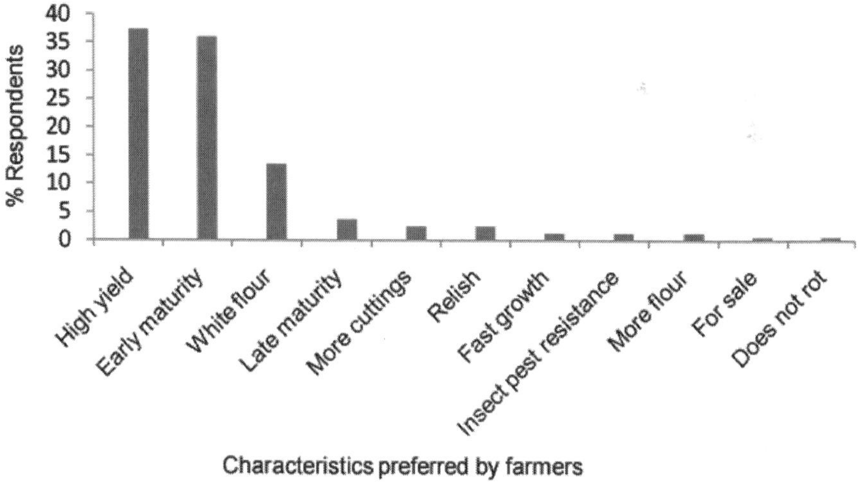

Figure 4. Characteristics of cassava preferred by the farmers.

3.9. Production and Marketing Constraints

During the FGD in each district, farmers highlighted a number of production and marketing constraints. The constraints included insect pests, lack of capital, late bulking genotypes, and lack of a market for cassava. In Mwense and Mansa districts, lack of capital was considered most important (Table 3), while in Samfya lack of planting material was ranked as the major constraint. However, for the structured survey 16% in Mansa district viewed lack of capital as a hindrance to cassava production as did 52% of the respondents in Mwense district. In contrast, 75% of the respondents in Samfya district regarded drought as an important production constraint.

For the marketing constraints, 41% of the respondents across all three districts considered distance to the market as a major constraint. About 32% of the respondents felt that lack of transport to move cassava produce to the market was a problem. Poor roads were considered by 21.7% of the respondents as being one of the marketing constraints. Very few famers (4.3%) regarded lack of a market for cassava produce as a problem.

3.10. Cropping System

In most of the farmers' fields, cassava was intercropped with maize (Zeamaydis), groundnut (Arachis hypogaea) or common bean (Phaseolus vulgaris). Field observations showed that most of the farmers grew maize for food security. Since the local cassava cultivars took long (2 to 3 years) to give appreciable yield, maize and other crops served as the immediate food source. A large proportion of the farmers (65.5%) in Mansa district intercropped cassava with either maize, common bean or sweet potato (Ipomoea batatus). About 27.6% cultivated sweet potato alone, while 6.9% practised both systems (intercropping and mono-cropping) (Figure 5).

Table 3. Ranking of cassava production constraints identified during the focus group discussions held in three districts of Luapula province, Zambia (2009).

Constraint	Farmers' score* per district		
	Mwense	Mansa	Samfya
Capital	1	1	7
Late maturing	-	4	-
Market	6	3	-
Insect pests and diseases	3	5	3
Cassava cuttings	4	-	1
Shortage of labour	-	2	6
New varieties	-	-	4
Extension information	-	-	2
Transport to market	5	6	-
Drying of cassava	2	8	-
Low soil fertility	-	-	5
Implements	-	7	-

* = denotes rank of constraint; 1 = most pressing constraint; 8 = least pressing constraint.

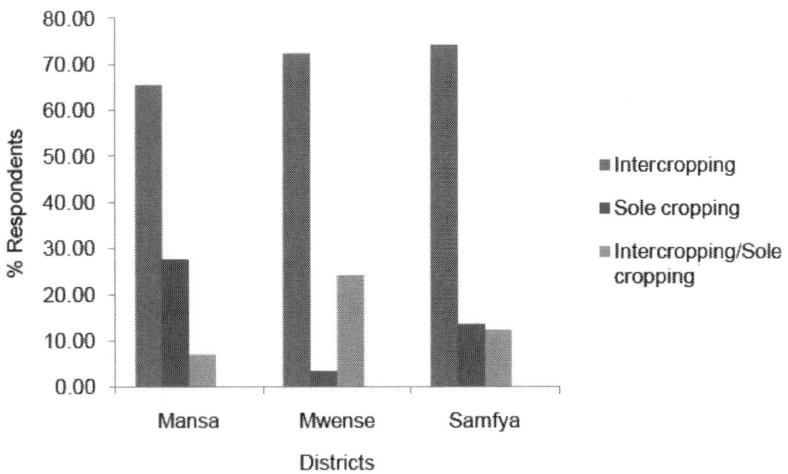

Figure 5. Cropping system practised by farmers in samfya, Mansa and mwense districts.

In Mwense district there were more farmers (72.4%) intercropping cassava than in Mansa district. However, few farmers (3.4%) grew cassava alone. In Samfya district, intercropping (83.9%) was the most practiced system followed by mono-cropping (9.7%). For farmers practicing cassava mono-cropping the reason given was that intercropping affected root development of cassava especially when harvesting sweet potato or groundnut.

4. DISCUSSION

The findings of the study revealed that the farmers did not know CMD as a disease in spite of its presence in most of the fields. Farmers associated CMD with mealybug infestation. The lack of knowledge of CMD by the farmers could be attributed to perceived damage to cassava plants. The susceptibility of improved varieties and landraces to CMD and other pests could be because the cultivars were not specifically bred for pest and disease resistance. The four officially released varieties (Mweru, Tanganyika, Kampolombo and Chila) were specifically breed for high yield although some are tolerant to pests and diseases. In addition, the failure by the majority of the farmers to recognise CMD as a disease necessitates effort on the part of researchers and extension officers to educate the farmers on diseases of cassava.

Technical support from the MACO and extension was almost non-existent in the three study areas. The few extension officers that participated in the group discussions were also ignorant of the disease. For the farmers to recognise CMD symptoms on leaves or differentiate between diseased or healthy cassava plants, requires properly trained extension officers both public and private. The level of education may have contributed to the poor understanding of the disease as most of the respondents were semi-literate. Although high CMD incidence (61.2%) was recorded, the disease was not recognised by the majority of farmers. This is contrary to studies in Kenya where 88% of the farmers recognised CMD as a disease of their crop [12] . Studies by

Muimba-Kanko- longo et al. [4] estimated 50% to 70% yield loss per year in Zambia. As it has long been observed [2] , the high levels of CMD incidence in the three districts surveyed require intervention by plant breeders to develop cassava varieties that are resistant to the disease.

Cassava mealybug (CMB) was regarded by all the farmers in the three districts as the major insect pest of cassava. However, no mention of whitefly (a vector of CMD) and cassava green mite was made. This could be due to the relatively small size of the insects. These observations are similar to findings reported by Poubom et al. [13] where farmers only mentioned large sized insects as the major constraints affecting cassava in West Africa. Similarly in Nigeria cassava green mite and CMB were considered to be diseases along cassava bacterial blight and Africa cassava mosaic virus disease while rodents, monkeys, and grasshoppers as pests [14] . The farmers indicated that termites and mealybugs caused the most damage to cassava plants. The importance of the mealybugs was recognised in all three districts as they are more easily noticed unlike whitefly.

There was no specific management of CMD by the minority of farmers who were aware of the disease. Although the farmers removed the affected top leaves (not specifically for CMD control), this clearly did not constitute effective management of CMD as detopping has been known to enhance symptom expression [15] .

The majority of the farmers obtained cassava planting materials from either their own fields or from fellow farmers. The exchange of cassava cuttings is not restricted to within farming communities but also occurs across districts. Consequently planting materials are often infected with viruses, a situation which contributes to the high incidence of CMD. This practice of exchanging or getting cassava cuttings for planting is one of the ways in which CMD is spread [16] . In Mansa district, few farmers obtained cuttings from each other, probably because they farm in close proximity to Mansa Regional Research Station, which has a cassava multiplication programme.

The potential for viruses to spread through the planting materials is exacerbated by the fact that none of the farmers was aware that planting materials may act as sources of further virus infection in the field.

Farmers grew a number of cultivars in their fields, most of them low yielding and susceptible to pests and diseases. Though the cultivars were susceptible, farmers, often unknowingly, protected the crop against diseases and pests with very little outside technical assistance. However, the crop protection methods applied were based on "trial and error" with little impact on addressing the CMD problem.

Growing of more than one crop and at different times of the year assured the farmers of food security. This is the underlying reason why intercropping is practised in the three districts. Most of the farmers intercropped cassava with annual crops e.g. bean, maize, groundnut, sweet potato and bambara nut (Vigna subterrania). Intercropping of maize and cassava has been demonstrated to improve productivity per unit land area [17] .

None of the farmers applied fertiliser although some of them acknowledged the low fertility levels of the cassava fields. This is similar to the practice of growing of cassava without fertiliser application among smallholder farmers in Democratic Republic of the Congo [18] .

Although most of the farmers acknowledged the presence of insect pests, not a single farmer used pesticides. This has implications on the management of pests especially where cultural practice cannot contain pest problem. The high cost of pesticides in Zambia and lack of technical know-how, probably restricted smallholder cassava farmers from using chemical control methods. A similar observation was made in Côte d'Ivoire [19] .

5. CONCLUSION

The study established that the farmers had little, if any, knowledge of CMD. High yield and early bulking were some of the traits preferred by the farmers. Most of the farmers prefer growing landraces because they have the desired attributes and are adapted to low-input conditions. Farmers' preferred traits need to be included in the selection criteria of cassava breeding programmes to meet the farmers' needs and expectations. The farmers identified a number of constraints with regard to cassava production, namely capital, labour and drought. Therefore, farmers' preferences, such as high yielding and early bulking traits which were widely mentioned by the farmers, have to be given attention in breeding programmes. The participation of farmers in breeding programmes, from early to advanced stages, will facilitate the adoption of new improved cultivars.

ACKNOWLEDGEMENTS

The authors would like to acknowledge Agriculture Green Revolution in Africa (AGRA) for the financial assistance towards the study. Special thanks go to the extension staff from Mansa, Samfya and Mwense districts for identifying and recruiting the farmers. We wish also to acknowledge all the farmers that participated in the study.

REFERENCES

1. Chitundu, M., Droppelmann, K. and Haggblade, S. (2006) A Value Chain Task Force Approach for Managing Private-Public Partnerships: Zambia's Task Force on Acceleration of Cassava Utilisation. Working Paper No. 21, Food Security Research Project.

2. Hahn, S.K., Terry, E.R., Leuschner, K. and Singh, T.P. (1981) Strategies of Cassava Improvement for Resistance to Major Economic Diseases and Pests in Africa. In: Terry, E.R., Oduro, K.A. and Caveness, F., Eds., Tropical Root Crops. Research Strategies for the 1980's. Proceeding of the 1st Triennial Root Crops Symposium of the International Society for Tropical Root Crops, Africa Branch, IITA, Ibadan, 25-28.

3. Thresh, J.M., Fargette, D. and Otim-Nape, G.W. (1994) Effects of African Cassava Mosaic Geminivirus on the Yield of Cassava. Tropical Science, 34, 26-42.

4. Muimba-Kankolongo, A., Chalwe, A., Sisupo, P. and Kang, M.S. (1997) Distribution, Prevalence and Outlook for Control of Cassava Mosaic Disease in Zambia. Roots, 4, 2-7.

5. FAOSTAT (2009) Food and Agriculture Organisation of the United Nations. Roma, Italy.http://faostat.fao.org/site/567/default.aspx

6. Agwu, A.E. and Anyaeche, C.L. (2007) Adoption of Improved Cassava Varieties in Six Rural Communities in Anambra State, Nigeria. African Journal of Biotechnology, 6, 89-98.

7. Manu-Aduening, J.A., Lamboll, R.I., Mensah, G.A., Lamptey, J.N., Moses, E., Dankyi, A.A. and Gibson, R.W. (2006) Development of Superior Cassava Cultivars in Ghana by Farmers and Scientists: The Process Adopted Outcomes and Contributions and Changed Roles of Different Stakeholders. Euphytica, 150, 47-61.

8. Joy, P. (1993) The Crisis of Farming Systems in Luapula Province, Zambia, Nordic. Journal of African Studies, 2, 118-140.

9. SPSS (2006) Statistical Programme for Social Sciences. SPSS for Windows. Release 2006, SPSS Inc

10. MACO (1991) Exploratory Soil Map of Zambia. Soil Survey Section, Zambia Agriculture Research Institute, Ministry of Agriculture and Cooperatives (MACO).

11. MACO (2008) Ministry of Agriculture and Cooperatives (MACO), 2003-2007 Crop Performance. Ridgeway, Lusaka.

12. Anderson, P.K. and Morales, F.J. (2005) Whitefly and Whitefly-Borne Viruses in the Tropics: Building a Knowledge Base for Global Action. Centro Internacional de Agricultura Tropical (CIAT), Cali, CO. 351 p. (CIAT Publication No. 341)

13. Poubom, C.F.N., Awah, E.T., Tchuanyo, M. and Tengoua, F. (2005) Farmers' Perceptions of Cassava Pests and Indigenous Control Methods in Cameroon. International Journal of Pest Management, 51, 157-164.

14. Nweke, F.I., Ugwu, B.O. and Dixon, A.G.O. (1996) Spread and Performance of Improved Cassava Varieties in Nigeria. Collaborative Study of Cassava in Africa (COSCA) Working Paper No.15.IITA, Ibadan, Nigeria.

15. Bellotti, A.C., Hershey, C.H. and Vargas, O. (1987) Recent Advances in Resistance to Insect and Mite Pests of Cassava. In: Hershey, C.H., Ed., Cassava Breeding: A Multidisciplinary Review, Proceedings of a Workshop Held in Philippines, CIAT, Cali, 4-7 March 1985, 117-146.

16. Ntawuruhunga, P., Legg, J., Okidi, J., Okao-Okuja, G., Tadu, G. and Remington, T. (2007) Southern Sudan Equatorial Region. Cassava Baseline Survey Technical Report, IITA, Ibadan, Nigeria, p: 65.

17. Adeniyan, O.N., Aluko, O.A., Olanipekun, S.O., Olasoji, J.O. and Aduramigba-Modupe, V.O. (2014) Growth and Yield Performance of Cassava/Maize Intercrop under Different Plant

Population Density of Maize. Journal of Agricultural Science, 6, 35.

18. Muengula-Manyi, M., Nkongolo, K.K., Bragard, C., Tshilenge-Djim, P., Winter, S. and Kalonji-Mbuyi, B. (2012) Incidence, Severity and Gravity of Cassava Mosaic Disease in Savannah Agro-Ecological Region of DR-Congo: Analysis of Agro-Environmental Factors. American Journal of Plant Sciences, 3, 512-519.

19. Nweke, F.I., Ngoram, K., Dixon, A.G.O., Ugwu, B.O. and Ajobo, O. (2000) Cassava Production and Processing in Cote D'Ivoire. Collaborative Study of Cassava in Africa (COSCA) Working Paper No. 23.IITA, Ibadan, Nigeria.

Chapter 14

Susceptibility to Citrus Canker Caused by Xanthomonas Axonopodis Pv. Citri Depends on The Nuclear Genome of The Host Plant

Hiroshi Shiotani[1] Terutaka Yoshioka[2] Masashi Yamamoto[3] Ryoji Matsumoto[4]

[1]*Kuchinotsu Citrus Research StationNational Institute of Fruit Tree ScienceNagasakiJapan*

[2]*National Institute of Fruit Tree ScienceIbarakiJapan*

[3]*Faculty of AgricultureKagoshima UniversityKagoshimaJapan*

[4]*Faculty of AgricultureSaga UniversitySagaJapan*

ABSTRACT

Susceptibility to *Xanthomonasaxonopodis* pv. *citri* of a citrus cybrid, in which the nuclear and cytoplasmic genomes were derived from *Citrussinensis* and *C. unshiu,* respectively, was evaluated. Bacterial growth in the leaves of the cybrid was similar to that in *C. sinensis* after pin-prick inoculation throughout the experiment, whereas growth was greater than that in *C.unshiu* from 8 days after inoculation. Lesion expansion and pustules that later developed from the lesions on the cybrid and on *C. sinensis* also appeared to be greater than those on *C. unshiu*. The incidence of citrus canker disease caused by the bacteria on the cybrid trees was in the field was equivalent to that on *C. sinensis* but was severer than on *C. unshiu*. These results indicate that the nuclear genome of the cybrid plays a critical role in susceptibility to citrus canker disease. However, the pathogenicity gene *pthA* of the bacteria is not likely to be involved in the difference in susceptibility to the bacteria between *C. unshiu* and *C. sinensis* because their susceptibility to a *pthA*-deficient mutant of the bacteria also differs.

Keywords: *Xanthomonas axonopodis* pv. *Citri* Citrus cybrid

INTRODUCTION

The causal agent of citrus canker disease, *Xanthomonasaxonopodis* pv. *citri*, has pathogenicity gene *pthA*, which belongs to the *avrBs3/pthA* (avirulence/pathogenicity) gene family and is required for full virulence on citrus plants (Kanamori and Tsuyumu 1998; Swarup et al. 1992). The transient expression of *pthA* in the leaves of *Citrus* species results in citrus canker symptoms involving cell hypertrophy, division, and death, which indicates that *PthA* functions inside the plant cell as a specific plant signal (Duan et al. 1999).

X. axonopodis pv. *citri* has a broad host range among members of the Rutaceae, although differences in field resistance exist among *Citrus* species (Gottwald and Graham 2000). The resistance of citrus plants has been considered as not due to active physiological resistance because all plants were highly susceptible after artificial inoculations (Brunings and Gabriel2003). However, another gene in the *avrBs3/pthA* family, *hssB3.0,* was recently isolated from*X. axonopodis* pv. *citri* and characterized as responsible for the host-specific suppression of virulence, which induces a defense response specifically on cultivars of *C. grandis* (Shiotani et al. 2000, 2007).

The *avrBs3/pthA* family is a group of highly conserved members prevalent in the genus*Xanthomonas*. The genes in this family confer two functions to the bacteria avirulence and virulence to the host plants with and without, respectively, the corresponding resistance gene (Gabriel 1999; Leach and White 1996). Both functions are highly dependent on the type III protein secretion system, which is encoded by the *hrp* gene cluster in the bacterial genome (Büttner and Bonus 2002). The functions of these gene products in the host plant have been deduced from their unique protein structure, which is composed of a central region of multiple, nearly identical repeated sequences of 34 amino acid residues, three nuclear localization signals (NLSs), and an acidic activation domain (AAD) in the carboxyl terminus (Ponciano et al. 2003). On the basis of the structure of the NLSs and the AAD, AvrBs3/PthA proteins are predicted to function in the plant nucleus (Szurek et al. 2001; Yang et al. 2000; Yang and Gabriel 1995). Mutations in the structures of both the NLSs and the AAD abolish the defense responses and reduce the ability of the pathogen to cause disease in various plant–pathogen interactions.

A citrus cybrid, which consists of the *C. sinensis* nuclear genome and the *C. unshiu* cytoplasmic genome, was produced to investigate cytoplasmic male sterility that probably originated from *C.unshiu* (Yamamoto and Kobayashi 1995). *C. unshiu* and *C. sinensis* have been considered to be moderately resistant and highly susceptible, respectively, to citrus canker disease (Matsumoto and Okudai 1990). The cytoplasm has been shown to be involved in cytoplasmic male sterility and vital plant functions such as photosynthesis, sugar and fatty acid metabolism, ATP production, and disease resistance (Kadowaki et al. 1990; Mazouz et al. 2002; Otani et al. 2002). Because there is little cytoplasmic information in citrus, this cybrid can be an ideal material for studying cytoplasmic involvement in resistance to citrus canker disease. We report here that the nuclei of host plants also play a critical role in this disease through experiments with this cybrid and its parent citrus plants.

MATERIALS AND METHODS

Plant materials

The citrus cybrid was produced previously by electrofusion between protoplasts derived from the nucellar embryo of *Citrus unshiu* MARC cv. Juman and a mesophyll cell of *C. sinensis*Osbeck cv. Washington naval orange (Yamamoto and Kobayashi 1995). Leaf morphology of the cybrid was similar to that of the mesophyll parent *C. sinensis*. Flower characteristics of the cybrid were almost identical to those of *C. sinensis* except for the number of stamens (Yamamoto et al. 2001). The cybrid consists of the *C. sinensis* nuclear genome and *C. unshiu*cytoplasmic genome. Scions of the cybrid and its parent plants used for the inoculation test were grafted onto potted two-year-old rootstocks *Poncirustrifoliata* Raf. (trifoliate orange). Disease was assessed in each of three trees, planted eight years before, in the orchard at the Kuchinotsu Citrus Research Station.

Bacterial strains

A strain KC21 of *Xanthomonasaxonopodis* pv. *citri* and its derivative mutant KC21T14, in which the pathogenicity gene *pthA-KC21* was disrupted by Tn5 insertion, were used in this study (Shiotani et al. 2007). The bacterial strains were cultured in yeast–peptone (YP) medium at 27°C. For solid medium, 1.5% agar was added to these media. The final concentration of the antibiotics incorporated into media was 50 µg per ml for kanamycin for selecting resistant strain KC21T14. The gene *pthA-KC21* was deposited in GenBank under accession number AB206388.

Inoculation test

The sensitivity of the cybrid and its parents plants to *X. axonopodis* pv. *citri* were evaluated after pin-prick inoculation of attached leaves with the bacterial strains as previously reported (Shiotani et al. 2000). Bacterial strains were grown on Wakimoto's potato semi-synthetic medium with 1.5% agar (PSA medium) (Wakimoto 1967) at 27°C for 1 day. Bacteria were suspended in 0.85% NaCl at a concentration of 10^8 cells/ml, and a drop of the suspension was placed on an attached, mature leaf of each citrus plant. Immediately after an insect pin (0.5 mm in diameter) was inserted into the leaves through the bacterial suspension, the inoculum drops were wiped off with sterile cotton. The inoculated plants were then grown in a greenhouse. The diameter of circular lesions, including cork tissue and water-soaked margin, was measured after inoculation. For observing bacterial growth *in planta*, the populations in the lesions at 0, 1, 2, 3, 4, 8, 16, and 32 days after pin-prick inoculation were determined after removing the lesion with a cork-borer. The excised tissue was macerated in 0.85% NaCl, and the suspension was plated on a semiselective XCSM medium (Shiotani et al. 2000).

Disease development in the orchard

Disease development was investigated in July 2003 on 20 shoots for each tree selected in order of lesion abundance, with the following index: of 0, no lesions; 1, 1–3 lesions; 3, 4–10 lesions; 5, 11–20 lesions; 7, more than 20 lesions. Disease severity (DS) was calculated from the sum of the index numbers (SI) as follows:

$$DS=SI/(20\times7)\times100DS=SI/(20\times7)\times100$$

RESULTS AND DISCUSSION

Susceptibility to citrus canker disease depends on the nuclear genome of the host plant

The sensitivity of the cybrid and its parents plant, *Citrus unshiu* (cv. Juman) and *C. sinensis* (cv. Washington navel orange), to *X. axonopodis* pv. *citri* was evaluated after pin-prick inoculation of attached leaves with the bacterial strain KC21. Water-soaked lesions were elicited on all citrus plants within 4 days after inoculation. Pustules later developed from the lesions; however, those on *C. unshiu* appear to have developed less than those on the cybrid and *C. sinensis*(Fig. 1). In all the plants tested, populations of strain KC21 increased similarly for the first 4 days after prick inoculation to approximately 3.4×10^5 CFU per lesion (Fig. 2a). After 4 days, the rate of multiplication on *C. unshiu* differed from those on the cybrid and *C. sinensis*. After 16 days, the bacterial populations decreased on *C. unshiu*, but not on the cybrid nor *C. sinensis*. No significant difference in the bacterial population was observed between the cybrid and *C.sinensis* throughout the experiment. These results indicate that the susceptibility to citrus canker disease in the leaf tissue of the cybrid plant, which is equivalent to that of *C. sinensis*, might depend on the nuclear genome of the species. Disease severity on the cybrid trees in the orchard was also similar to that on *C. sinensis* trees (Table 1). Citrus cultivars and species with greater frequency, size, and duration of leaf flushes are more susceptible to citrus canker in the field than are less vigorous cultivars or those with more rapidly maturing foliage (Gottwald et al.1993). Because no difference in vigor has been observed between the cybrid and *C. sinensis*trees, the identical susceptibility of the cybrid leaf tissue and *C. sinensis* to *X. axonopodis* pv.*citri* appeared to be related directly to the field susceptibility of the cybrid trees to citrus canker. Cytoplasm can contribute disease resistance (Mazouz et al. 2002; Ohtani et al. 2002); however, the moderate resistance to citrus canker of *C. unshiu* might be independent of its cytoplasmic genome.

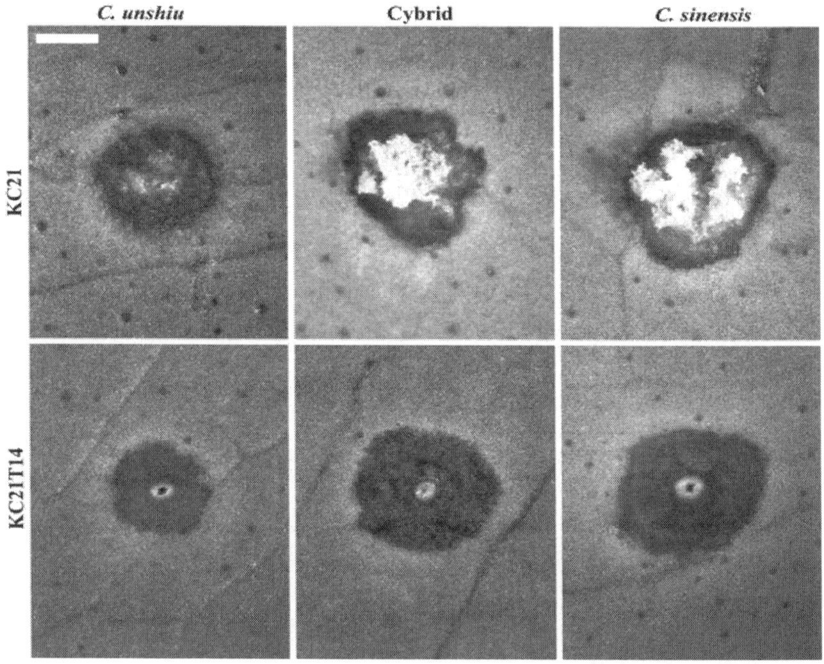

Figure 1. Canker symptoms on abaxial leaf surface of *Citrussinensis, C. unshiu,* and the cybrid between them. Symptoms developed 40 days after prick inoculation with strain KC21 and its derivative *pthA-KC21*-deficient mutant KC21T14 of *Xanthomonas axonopodis* pv. *citri*. *Bar* represents 1 mm

Table 1. Severity of citrus canker disease on *Citrus* species and their cybrid

Plant	Disease severity[a]
C. unshiu	3.1 ± 0.1b
C. sinensis	4.8 ± 0.5a
Cybrid	5.0 ± 0.4a

[a] Disease severity on each plant was investigated in July 2003, on 20 shoots for each tree selected in order of lesion abundance, index: of 0, no lesions; 1, 1–3 lesions; 3, 4–10 lesions; 5, 11–20 lesions; 7, more than 20 lesions. The mean value and standard error of three trees is presented. Those followed by unlike letters differ significantly at $\alpha = 0.01$ according to the Tukey–Kramer honestly significant difference test

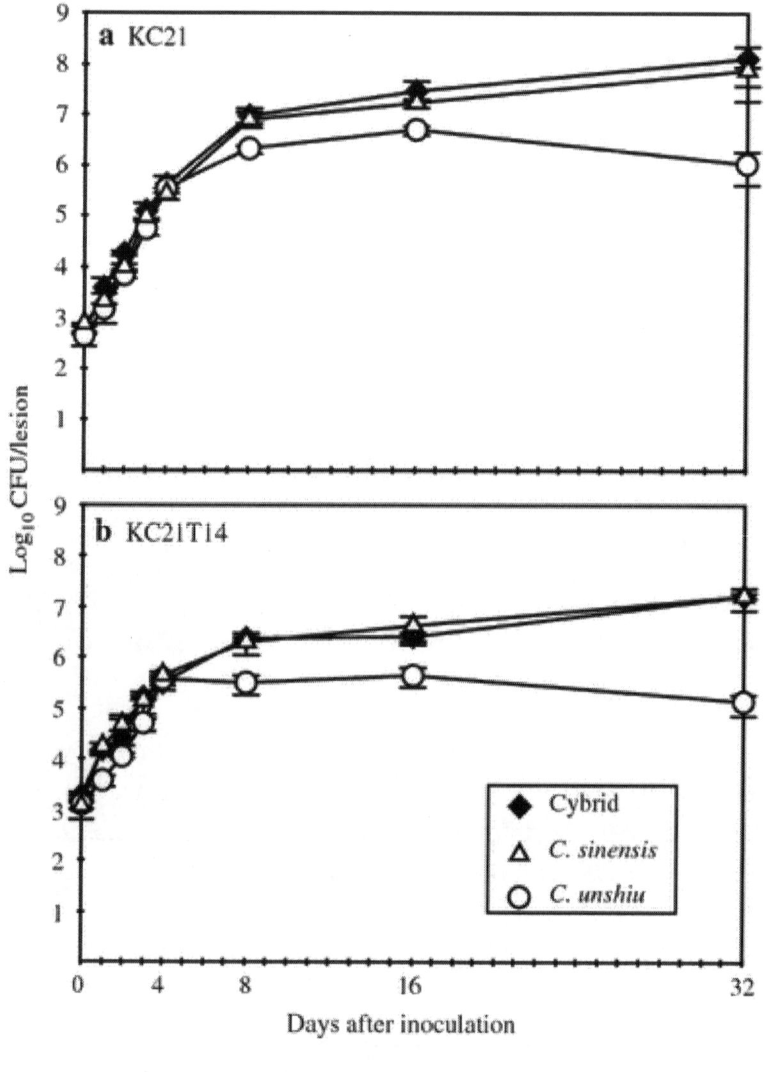

Figure 2. Time course of bacterial growth in the leaves of citrus plants. Leaves of *Citrus sinensis*, *C. unshiu*, and their cybrid were prick inoculated with strain KC21 (**a**) of *Xanthomonasaxonopodis* pv. *citri* and its derivative mutant KC21T14 (**b**), in which the pathogenicity gene *pthA-KC21* was disrupted by Tn5 insertion. Leaves were sampled for 32 days after inoculation. The values are means from five repetitions, and *vertical bars* represent the standard error at each sampling time

Different rates of lesion expansion between on leaves of the cybrid and *C. sinensis* were observed (Table 2). This observation suggests that this difference depends not on their nuclear but on the cytoplasmic genomes. However, this difference in lesion expansion does not mean that the susceptibility to citrus canker differs between these plants because bacterial growth on these two was similar for 32 days after inoculation (Fig. 2a).

Table 2. Lesion diameters on leaves of *Citrus* species and their cybrid 40 d after prick-inoculation with either wild-type strain KC21 and its Tn5 mutant KC21T14 of *Xanthomonas axonopodis* pv. *Citri*

Plant	Lesion diameter (mm)[a]	
	KC21	KC21T14
C. unshiu	2.63 ± 0.06c	1.69 ± 0.27b
C. sinensis	3.56 ± 0.05a	2.34 ± 0.20a
Cybrid	3.07 ± 0.07b	2.23 ± 0.13a

[a] Each value represents the mean diameter (mm) and standard error of 16 lesions for each strain on attached leaves. Those followed by unlike letters differ significantly at $\alpha = 0.01$ according to the Tukey–Kramer honestly significant difference test

Pathogenicity gene *pthA* of *X. axonopodis* pv. *citri* is independent of differences in susceptibility of *C. unshiu* and *C. sinensis* to the bacteria

The pathogenicity gene *pthA*, essential for *X. axonopodis* pv. *citri* to cause citrus canker disease, confers the ability to rupture the epidermis and elicit cankers on citrus. Thus, *PthA* may be involved in the difference in pustule development observed between *C. sinensis* and *C. unshiu*, in which nuclear genomes directly affect the susceptibility to citrus canker disease. To determine the involvement of *pthA* in the different pustule development, we produced a mutant KC21T14, derived from a strain KC21 of *X. axonopodis* pv. *citri*, in which pathogenicity gene*pthA-KC21* was disrupted by insertion of Tn5, was used to prick inoculate the leaves of *C.unshiu* and *C. sinensis*. The mutant KC21T14 elicited water-soaked lesions on their leaves, but no pustules were observed on the lesions throughout the experiment (Fig. 1). Also, the mutant grew significantly less in the lesions on both *Citrus* species than did the wild strain KC21 from 8 days after inoculation (Fig. 2b). Although lesions on the cybrid and *C. sinensis* expanded at different rates after inoculation with a wild strain KC21, the *pthA-KC21*-deficient mutant KC21T14 induced lesions of the same size on both (Table 2). This suggests that *pthA-KC21*might be associated in the different lesion expansion.

The differences in bacterial growth and lesion expansion were still observed between on *C.unshiu* and *C. sinensis* even after the inoculation with the mutant KC21T14 (Fig. 2b and Table 2). These results indicates that the pathogenicity gene *pthA-KC21* in the strain KC21 is independent of the difference in susceptibility to the bacteria between these *Citrus* species although this gene contributes to eliciting the distinctive canker symptoms (Brunings and Gabriel 2003). In addition, the bacterial population and lesion expansion on the cybrid after inoculation with mutant KC21T14 was the same as those on *C. sinensis* not on *C. unshiu*. These results strongly suggest that the nuclear genome of *C. sinensis* confers greater susceptibility to*X. axonopodis* pv. *citri*.

Strain KC21 harbors three genes in the *avrBs3/pthA* family other than *pthA-KC21*: *pB3.1*,*pB3.7*, and *hssB3.0* (Shiotani et al. 2007). Of these, *hssB3.0* is involved in the host-specific suppression of virulence, so that a defense response is elicited in *C. grandis*. Because all these family genes encode NLS sequences that are identical to those of *pthA*, their protein products could also be transferred into the nucleus of host plants. However, there is no evidence that any of the *avrBs3/pthA* family genes are involved in the moderate resistance of *C. unshiu* to *X.axonopodis* pv. *citri*. According to the gene-for-gene theory (Gabriel 1999), the gene product of*hssB3.0* might be recognized by a cognate resistance gene. Further studies will need to be done to elucidate the resistance gene in *Citrus* genus.

REFERENCES

1. Brunings AM, Gabriel DW (2003) *Xanthomonas citri*: breaking the surface. Mol Plant Pathol 4:141–157

2. Büttner D, Bonas U (2002) Getting across-bacterial type III effector proteins on their way to the plant cell. EMBO J 21:5313–5322

3. Duan YP, Castañeda A, Zhao G, Erdos G, Gabriel DW (1999) Expression of a single, host-specific, bacterial pathogenicity gene in plant cells elicits division, enlargement, and cell death. Mol Plant Microbe Interact 12:556–560

4. Gabriel DW (1999) The *Xanthomonas avr/pth* gene family. In: Stacey G, Keen NT (ed) Plant-microbe interactions vol. 4. APS Press, St. Paul, pp 39–55

5. Gottwald TR, Graham JH, Civerolo EL, Barrett HC, Hearn CJ (1993) Differential host range reaction of citrus and citrus relatives to citrus canker and citrus bacterial spot determined by leaf mesophyll susceptibility. Plant Dis 77:1004–1009

6. Gottwald TR, Graham JH (2000) Canker. In: Timmer LW, Garnsey SM, Graham JH (eds) Compendium of citrus diseases, 2nd edn. APS Press, St. Paul, pp 5–7

7. Kanamori H, Tsuyumu S (1998) Comparison of nucleotide sequences of canker-forming and non-canker-forming *pthA* homologues in *Xanthomonas campestris* pv. *citri*. Ann Phytopathol Soc Jpn 64:462–470

8. Kadowaki K, Suzuki T, Kazama S (1990) A chimeric gene containing the 5' portion of *atp6* is associated with cytoplasmic male-sterility. Mol Gen Genet 224:10–16

9. Leach JE, White FF (1996) Bacterial avirulence genes. Annu Rev Phytopathol 34:153–179

10. Matsumoto R, Okudai N (1990) Inheritance of resistance to bacterial canker disease in citrus (in Japanese). J Jpn Soc Hortic Sci 59:9–14

11. Mazouz H, Saadaoui EM, Jlibene M, Bouami FE (2002) Evidence of cytoplasm contributing resistance to specific groups of virulence of *Mycosphaerella graminicola* (Fuckel) Shroeter in two bread wheat genotypes (*Triticum aestivum* L.). Field Crop Res 79:197–206

12. Ohtani K, Yamamoto H, Akimitsu K (2002) Sensitivity to *Alternaria alternata* toxin in citrus because of altered mitochondrial RNA processing. Proc Natl Acad Sci USA 99:2439–2444

13. Ponciano G, Ishihara H, Tsuyumu S, Leach JE (2003) Bacterial effectors in plant disease and defense: keys to durable resistance? Plant Dis 87:1272–1282

14. Shiotani H, Fujikawa T, Ishihara H, Tsuyumu S, Ozaki K (2007) A *pthA* homolog from*Xanthomonas axonopodis* pv. *citri* responsible for host-specific suppression of virulence. J Bacteriol 189:3271–3279

15. Shiotani H, Ozaki K, Tsuyumu S (2000) Pathogenic interactions between *Xanthomonasaxonopodis* pv. *citri* and cultivars of Pummelo (*Citrus grandis*). Phytopathology 90:1383–1389

16. Swarup S, Yang Y, Kingsley MT, Gabriel DW (1992) A *Xanthomonas citri* pathogenicity gene*pthA*, pleiotropically encodes gratuitous avirulence on nonhosts. Mol Plant Microbe Interact 5:204–213

17. Szurek B, Marois E, Bonus U, Van den Ackerveken G (2001) Eukaryotic features of the*Xanthomonas* type III effector AvrBs3: protein domains involved in transcriptional activation and the interaction with nuclear import receptors from pepper. Plant J 26:523–534PubMedCrossRef

18. Wakimoto S (1967) Some characteristics of citrus canker bacteria, *Xanthomonas citri* (Hasse) Dowson, and the related phages isolated from Japan. Ann Phytopathol Soc Jpn 33:301–310

19. Yamamoto M, Kobayashi S (1995) A cybrid plant produced by electrofusion between *Citrus unshiu* (Satsuma mandarin) and *C. sinensis* (sweet orange). Plant Tissue Cult Lett 12:131–137

20. Yamamoto M, Kobayashi S, Yoshioka T, Matsumoto R (2001) Cybridization in *Citrus unshiu*Marc. (satsuma mandarin) and *C. sinensis* (L.) Osbeck (sweet orange). In: Nagata T, Bajaj YPS (eds) Somatic hybridization in crop improvement II. Biotechnology in agriculture and forestry vol. 49. Springer, Berlin pp 124–138

21. Yang Y, Gabriel DW (1995) *Xanthomonas* avirulence/pathogenicity gene family encodes functional plant nuclear targeting signals. Mol Plant Microbe Interact 8:627–631

22. Yang B, Zhu W, Johnson LB, White FF (2000) The virulence factor AvrXa7 of *Xanthomonasoryzae* pv. *oryzae* is a type III secretion pathway-dependent nuclear-localized double-stranded DNA-binding protein. Proc Natl Acad Sci USA 97:9807–9812

Index